INVESTIGATING MEDIUMS

Julie Beischel, PhD

Investigating
Mediums

A Windbridge Institute Collection

Julie Beischel, PhD

Praise for Julie Beischel, PhD, and the Windbridge Institute...

"Julie Beischel and her team at the Windbridge Institute have joined a unique group of scientists over the past century who have studied mediums. Beischel's team has taken on the challenge in a serious and rigorous way, and the results of their efforts are intriguing."
—Dean Radin, PhD, Senior Scientist, IONS

"Dr. Julie Beischel is a courageous, innovative pioneer who has brought immense respectability to a controversial area. Her contribution is huge, because throughout human history the terror of death's finality has caused more suffering than all the physical diseases combined."
—Larry Dossey, MD, Author of *One Mind*

"The research that is currently being done at the Windbridge Institute is the best research being done in the field of parapsychology."
—John Kruth, Exec. Director, Rhine Research Center

"Beischel herself has played a major role in strengthening the methods used in this field and her research exemplifies her rigor and sophistication."
—Daryl Bem, PhD, Professor Emeritus of Psychology, Cornell University

"Probably no one has done as much detailed scientific research as Dr. Beischel in such an original way. Her results are unquestionably authentic and her unique direction unparalleled."
–Samadhi Longo-Disse, M.Div, STM, PhD, Grief Counseling Professional

"The Windbridge Institute is unique in the world today and is incredibly important in giving credibility to mediumship."
—Victor Zammit, co-author of *A Lawyer Presents the Evidence for the Afterlife*

"Dr. Beischel's novel approach to mediumship research involves treating mediums as human beings who are able to report reliably about their inner subjective experience while giving readings. This approach has provided the fatal blow to the remains of the so-called 'superpsi' hypothesis which has plagued mediumship research since the time of William James."
—Neal Grossman, PhD, Professor Emeritus of Philosophy, University of Illinois at Chicago

"It has been only in the last 15 or so years that a few courageous scientists have risked showing any interest in studying mediumship and picking up where [the] esteemed researchers of yesteryear left off. Dr. Beischel is one of those few."
—Michael Tymn, author of *The Afterlife Revealed: What Happens After We Die*

"Beischel is to be congratulated for her dedication to understanding a type of human experience that, for largely ideological reasons, science as a whole has shunned."
—Robert McLuhan, author of *Randi's Prize, What Sceptics Say about the Paranormal, Why They Are Wrong and Why It Matters*

Julie Beischel, PhD

The Windbridge Institute, LLC
1517 N. Wilmot Rd. #254
Tucson, AZ 85712
http://www.windbridge.org/
info@windbridge.org

Text copyright © 2015 by Julie Beischel, PhD

All Rights Reserved

The names of some individuals and locations in this book have been changed. The information in this publication is provided "as is" without warranty of any kind, either express or implied, including but not limited to the implied warranties of merchantability and fitness for a particular purpose. Under no circumstances shall the Windbridge Institute, LLC, nor any party involved in creating, producing, or delivering this publication be liable for any damages whatsoever including direct, indirect, incidental, consequential, loss of business profits, or special damages. Always consult with your physician or other qualified healthcare provider when seeking treatment options.

*For Mark,
my smizmar.*

Julie Beischel, PhD

Investigating Mediums
Contents

- ∾1∾ An Introduction to the Collection
- ∾7∾ Mental Mediumship Research
- ∾17∾ Assisted after-death communication: A self-prescribed treatment for grief
- ∾27∾ Mediums' Experiences of the Afterlife
- ∾31∾ Book review: *The Truth About Grief*
- ∾37∾ Animal Discarnates
- ∾59∾ *Among Mediums*
- ∾173∾ *Meaningful Messages*
- ∾217∾ *From the Mouths of Mediums Vol. 1*
- ∾325∾ Negligence or Reality?
- ∾331∾ Further Resources
- ∾335∾ About the Author

Julie Beischel, PhD

An Introduction to the Collection

*Against other things
it is possible to obtain security,
but when it comes to death,
we all live in a city without walls.*
—Epicurus

Death will get around to all of us sooner or later.

One way we scientists address this inevitability is researching what happens to the mind, the soul, the self, the 'I' after the body dies. And one way we do that is by studying mediums, people who experience regular communication with the deceased.

There are only a handful of us performing mediumship research today anywhere in the world. To the best of my knowledge, I am the only scientist in the US performing empirical research with mediums full-time. And I've been doing so for more than 12 years.

This collection includes three of my previously published e-books as well as some articles:

"Mental Mediumship Research at the Windbridge Institute" (2015) is an article from the official journal of the World Institute for Scientific Exploration (WISE) describing our three-tiered approach to studying the accuracy and specificity of the information secular, American mediums report as well as their experiences, psychology, and physiology and the potential social applications of mediumship readings.

"Assisted after-death communication: A self-prescribed treatment for grief" (2014) is an extended abstract based on a presentation given at the 2013 conference of the American Center for the Integration of Spiritually Transformative Experiences (ACISTE, pronounced 'assist'). It was originally published in 2014 the *Journal of Near-Death Studies*.

"Certified Mediums' Experiences with the Afterlife," "Recommended Reading: *The Truth About Grief*," and "Survival of Consciousness: Animal Discarnates" are all articles from *Winds of Change*, the Windbridge Institute newsletter.

Among Mediums: A Scientist's Quest for Answers (2013) discusses how I went from interdisciplinary training in physiology and a PhD in pharmacology and toxicology with a minor in microbiology and immunology to pursuing rigorous scientific research with mediums full-time and the answers I discovered along the way.

Meaningful Messages: Making the Most of Your Mediumship Reading (2013) provides 10 helpful hints to keep in mind as you prepare for, experience, and reflect on a reading from a medium. Having realistic expectations and a clear understanding of the process will make the whole experience go more smoothly for the (at least) three people involved: you, the medium, and your deceased loved one.

Not being a medium myself, I enlisted the help of actual practicing mediums with decades of experience to share what's it's like to be a medium. In *From the Mouths of Mediums Vol. 1: Experiencing Communication* (2014), 13 mediums on our research team discuss how they experience communication from the deceased; suggestions for how you can experience communication; and why it might be that you have not heard from your loved one. The mediums who contributed were Ankhasha Amenti, Traci Bray, Dave Campbell, Carrie D. Cox, Joanne Gerber, Daria Justyn, Nancy Marlowe, Sarah McLean, Tracy Lee Nash, Troy Parkinson, Ginger Quinlan, Eliza Rey, and Kim Russo (The Happy Medium).

And finally, "Academic Negligence or Economic Reality?" is a *Winds of Change* article describing why more scientists aren't researching the survival of consciousness, a topic that affects every single person ever.

Notes on the Collection

There will be some overlap in the content from these sources (as there are only so many ways for me to say what I do, how I do it, and what I've found).

The formatting of the references, etc., is not consistent between sources and reflects the format of the originals. Also, my rampant use of bullets (much to my husband's chagrin) is not supported in this format and simple, indented, or lettered lists are used as a replacement here. (I'm quite confident no meaning was lost along with the disposed bullets.)

I originally wrote each e-book keeping in mind the philosophy that brevity is the key to good communication. We're all busy and I could convey the important information without taking up a lot of your (and, let's be honest, my) time. Also, by offering the three short e-books without print versions, we were able to keep the prices low making them more accessible to more people. However, we have received many requests for print versions of the e-books but printing any one book would not be cost-effective (who wants to pay 12 bucks for 50 pages?). A collection of all of them together is long enough, though, that it made printing a viable option.

It was almost painful trying not to perfect and/or update every sentence in this collection but I would have been at that forever and would still not be satisfied. Thus, the content here reflects the original publications.

My canine partner, Moose, who I lovingly acknowledge throughout these sources passed away in March, 2015. She was 15 years old. I miss her terribly every day.

Other journal articles, research briefs, and media are available on the Windbridge Institute website at: http://www.windbridge.org/publications/

Julie Beischel, PhD

Mental Mediumship Research at the Windbridge Institute

[This article was originally published in the Summer 2015 edition of the WISE Journal: The official journal of the World Institute for Scientific Exploration (WISE), *Volume 4, No. 2, pp. 41-44.]*

One way to scientifically examine the theory that consciousness survives physical death (the survival of consciousness hypothesis or simply 'survival') is through the investigation of the phenomenon of mediumship. Although it is theoretically possible for anyone to experience communication from the deceased, a medium is someone who has this experience regularly, reliably, and often on-demand. There are two main categories of mediumship: physical and mental.

During physical mediumship, physical effects including materializations and table-tipping are produced. The purpose of mental mediumship is to convey messages from deceased people or animals (called discarnates) to their living loved ones (called sitters) during a specific event (called a reading). (Psychics, on the other hand, convey information about people, events, places, or times unknown to them but not about the deceased. It is often said that all mediums are psychic but not all psychics are mediums.) At the Windbridge

Institute, we have been performing research since 2008 with a team of mental mediums whose abilities have been demonstrated under controlled conditions. This allows us to study the phenomenon with participants who are able to effectively and repeatedly provide readings during various research protocols.

Before participating in mediumship research at the Windbridge Institute, each prospective medium is screened over several months using an intensive eight-step screening procedure (1). Upon successful completion of the eight steps, the medium is termed a Windbridge Certified Research Medium (WCRM). Each WCRM agrees to donate a minimum of four hours per month to assist in various aspects of the research, uphold a code of spiritual ethics, embrace a strong commitment to the values of scientific mediumship research, and abide by specific Windbridge Institute standards of conduct.

At the Windbridge Institute, studies with mediums are performed as part of three research programs: Information, Operation, and Application.

Information Research Program

This program examines the abilities of mediums to report accurate and specific information about discarnates using anomalous information reception (AIR); that is, without any prior knowledge or feedback and without using deceptive or fraudulent means.

These experiments employ a more-than-double-blinded research reading protocol (1) which involves five levels of blinding that account for fraud, cueing, cold reading including generality, rater bias, and precognition as explanations for the source of the mediums' information:

(a) the WCRM is blinded to information about the sitter and the discarnate before, during, and after the phone reading;

(b) the sitters are blinded to the origin of the readings during scoring; they do not hear the readings as they occur and receive one target and one decoy reading to score;

(c) the experimenter who consents and trains the sitters (Experimenter 1) is blinded to which mediums read which sitters and which readings were intended for which sitters;

(d) the experimenter who interacts with the mediums during the phone readings and formats the readings into item lists (Experimenter 2) is blinded to information about the sitters and the discarnates beyond the discarnates' first names;

(e) the experimenter who interacts with the sitters during scoring (i.e., e-mails and receives by e-mail the blinded readings; Experimenter 3) is blinded to information about the discarnates, to which medium performed which readings, and to which readings were intended for which sitters.

The resulting accuracy data collected between 2009 and 2013 included 58 scored readings performed by 20 WCRMs and scored by 116 sitters (4). This study (AIRII) served to replicate and extend data collected during a previous study (7) and provides statistical evidence that mediums can report accurate and specific information about the deceased under fully blinded conditions that address all normal, sensory explanations for how they could have acquired that information.

Operation Research Program

This program examines the unique experiences (phenomenology), psychology, and physiology of mediums.

Phenomenology

Because evidence for AIR alone cannot differentiate between two major parapsychological theories regarding the source of mediums' information [i.e., survival psi (mediums are receiving information from discarnates) and somatic psi (mediums are using clairvoyance, precognition, and/or telepathy with the living to retrieve information about discarnates)], investigating mediums' experiences addresses the source of their information. Taken together, (a) mediums' spontaneous reports that they are communicating directly with the deceased as well as (b) their alleged ability to differentiate between survival psi and somatic psi (9), and (c) experimental evidence that the two experiences are in fact different under controlled conditions (6), the explanation best supported by the data is survival psi; that is, it seems that at least some mediums are experiencing information from the dead.

Psychology
Mediums, as a group, tend to have personality types that are intuitive (able to make connections among diverse information) and feeling (concerned with harmony between people) (2). They also score statistically above average on instruments assessing conscientiousness, openness to experiences, and transliminality (3).

Physiology
During an EEG study, six WCRMs were asked to consciously enter into four different mental states including communication with the deceased. The findings (8) suggest that the experience of communication may be a distinct mental state that is not consistent with brain activity during ordinary recollection, perception, or imagination. Thus, the claims that all mediums are fabricating information or recalling facts previously acquired about discarnates are not supported.

When asked about their health challenges, mediums report an incidence of migraine headaches and autoimmune disorders statistically higher than would be expected by chance. A recent study examining possible hematological and/or physiological correlates of the mediumistic state in five WCRMs (3) found no detectable changes. Because mediums also report a higher than normal incidence of childhood trauma and a general link between childhood trauma and adult disease has been established, we have proposed that the disease prevalence in mediums may be linked to trauma (3). We are currently developing a general survey in an attempt identify potentially unique demographic, cognitive, psychological, physiological, familial, cultural, and phenomenological characteristics of self-identifying mediums in the US to further examine this link in a wider population.

Application Research Program

This research program attempts to address the potential social applications of mediumship. At this time, our main focus is on the potential therapeutic effects on grief of assisted after-death communication during readings with mediums (5) though we are also interested in applications in law enforcement.

References

1. Beischel, J. (2007). Contemporary methods used in laboratory-based mediumship research. *Journal of Parapsychology*, *71*, 37-68.

2. Beischel, J. (2008, March). *Distinctive traits of those who 'speak with spirit': analysis of claimant mediums unique personality characteristics using three standard questionnaires*. Presented at the 28th Annual Society for the Anthropology of Consciousness Spring Conference, New Haven, Connecticut.

3. Beischel, J. (2015, May). *Assessing hematological and psychophysiological correlates of anomalous information reception in mediums*. 34th Annual Meeting of the Society for Scientific Exploration, Rockville, Maryland.

4. Beischel, J., Boccuzzi, M., Biuso, M., & Rock, A. J. (2015). Anomalous information reception by research mediums under blinded conditions II: Replication and extension. *EXPLORE: Journal of Science & Healing*, *11*(2), 136-142. doi:10.1016/j.explore.2015.01.001

5. Beischel, J., Mosher, C. & Boccuzzi, M. (2014-2015). The possible effects on bereavement of assisted after-death communication during readings with psychic mediums: A continuing bonds perspective. *Omega: Journal of Death and Dying*, *70*(2), 169-194. doi: 10.2190/OM.70.2.b

6. Beischel, J., Rock, A., & Boccuzzi, M. (2013, June). *The source of mediums information: A quantitative phenomenological analysis*. Presented at the 32nd Annual Meeting of the Society for Scientific Exploration, Dearborn, Michigan.

7. Beischel, J., & Schwartz, G.E. (2007). Anomalous information reception by research mediums demonstrated using a novel triple-blind protocol. *EXPLORE: Journal of Science & Healing, 3*, 23-27.

8. Delorme, A., Beischel. J., Michel, L., Boccuzzi, M., Radin, D., & Mills, P. J. (2013). Electrocortical activity associated with subjective communication with the deceased. *Frontiers in Psychology, 4*: 834. doi: 10.3389/fpsyg.2013.00834

9. Rock, A. J., Beischel, J., & Cott, C. C. (2009). Psi vs. survival: A qualitative investigation of mediums' phenomenology comparing psychic readings and ostensible communication with the deceased. *Transpersonal Psychology Review, 13*, 76-89.

INVESTIGATING MEDIUMS

Julie Beischel, PhD

Assisted After-Death Communication: A Self-Prescribed Treatment for Grief

[This extended abstract based on a presentation given at the 2013 conference of the American Center for the Integration of Spiritually Transformative Experiences (ACISTE, pronounced 'assist') was originally published in 2014 the Journal of Near-Death Studies. *The 'in press references have been updated in this version. A pdf version can be downloaded at http://www.windbridge.org/publications]*

citation: Beischel, J. (2014). Assisted after-death communication: A self-prescribed treatment for grief [Extended abstract]. *Journal of Near-Death Studies*, *32*, 161–165.

Estimates reveal that almost a third of American adults have had contact with the deceased (LaGrand, 2005; Newport & Strausberg, 2001). These spontaneous after-death communication (ADC) experiences include sensing the presence of the deceased; visual, olfactory, tactile, and auditory phenomena; powerful dreams; hearing meaningfully timed music associated with the deceased; lost-things-found; communication through electric devices; symbolic messages; synchronicities; and other phenomena seemingly unexplainable through the prevailing Western materialist, reductionist worldview (e.g., Barbato et al., 1999; Conant, 1996; Daggett, 2005; Drewry, 2003; Haraldsson, 1988; Houck, 2005; Klugman, 2006; LaGrand, 2005; Normand, Silverman, & Nickman, 1996; Nowatzki & Grant Kalischuk, 2009; Sanger, 2009; Sormanti & August, 1997). Many experts have asserted that these experiences are ubiquitous: They are not limited to specific socioeconomic groups or types of deaths or to particular times after the death (Dannenbaum & Kinnier, 2009; Houck, 2005; LaGrand, 2005). They appear to be a natural part of the grieving process (e.g., Barbato et al., 1999; Klugman, 2006; LaGrand, 2005). Several researchers have specifically examined the effects of spontaneous ADC experiences on the bereaved and found that they fulfill specific psychological needs and positively impact the grieving process (e.g., Conant, 1996; Drewry, 2003; LaGrand, 2005; Parker, 2005; Nowatzki & Grant Kalischuk, 2009; Sormanti & August, 1997; reviewed in Krippner, 2006). Similarly, induced ADC experiences have also been shown to positively affect the resolution of grief (Botkin, 2000; Hastings et al., 2002).

This extensive body of research implies a potential for similar effects after assisted ADCs, the experience of hearing from deceased loved ones during readings with mediums. Mediums, individuals who regularly experience communication with the deceased, are often stereotyped as charlatans based on the historical prevalence of such characters in their profession. However, the phenomenon of mediumship has been examined scientifically since the 1880s by researchers including William James and Oliver Lodge. Modern mediumship research includes studies of the accuracy of mediums' statements under controlled conditions (e.g., Beischel, Boccuzzi, Biuso, & Rock, submitted; Beischel & Schwartz, 2007; Kelly & Arcangel, 2011; Roy & Robertson, 2004) as well as examinations of their phenomenology (e.g., Beischel & Rock, 2009), psychology (e.g., Roxburgh & Roe, 2011), physiology (e.g., Beischel, 2013), and neurobiology (e.g., Delorme et al., 2013; Hageman et al., 2010). The three types of content commonly found in mediumship readings include information identifying of the deceased, information about events in the life of the bereaved that have occurred since the death, and direct messages. In general, this information assists the grieving in recognizing that their relationships with the deceased still exist, thus falling within the continuing bonds model of grief (Klass, Silverman, & Nickman, 1996).

Many bereaved individuals are presently receiving readings from mediums, though little is known about the effects of this self-prescribed treatment option. For example, assisted ADCs may be more preferable to people fearful of unexpected contact from the deceased and may also be useful for those who desire contact but have not yet experienced it. Work with a licensed mental health professional (MHP) in combination with a mediumship reading may be beneficial in addressing the acute grief experiences of the bereaved. In those cases, MHPs may wish to convey to clients that deciding to receive a reading should be done with significant forethought and intention, and they should consider providing resources to clients to create appropriate expectations of the mediumship process.

Initial case studies and exploratory survey data demonstrate profound effects on grief after readings with mediums (Beischel, Mosher, & Boccuzzi, 2014-2015), though no systematic studies have been performed. The Bereavement And Mediumship (BAM) Study is a randomized clinical trial examining the potential benefits of personal mediumship readings. Its aim is to effectively determine if receiving mediumship readings is helpful, harmful, or neither.

References

Barbato, M., Blunden, C ., Reid, K ., Irwin, H., & Rodriquez, P. (1999). Parapsychological phenomena near the time of death. *Journal of Palliative Care*, *15*, 30–38.

Beischel, J. (2013). *Among mediums: A scientist's quest for answers*. Retrieved from http://www.amazon.com/dp/B00B1MZMHM/

Beischel, J., Boccuzzi, M., Biuso, M., & Rock, A. (submitted). Anomalous information reception by research mediums under blinded conditions II: Replication and extension. *Explore: Journal of Science and Healing*.

Beischel, J., Mosher, C. & Boccuzzi, M. (2014-2015). The possible effects on bereavement of assisted after-death communication during readings with psychic mediums: A continuing bonds perspective. Omega: Journal of Death and Dying, 70(2), 169-194. doi: 10.2190/OM.70.2.b

Beischel, J., & Rock, A. J. (2009). Addressing the survival vs. psi debate through process-focused mediumship research. Journal of Parapsychology, 73, 71–90.

Beischel, J., & Schwartz, G.E. (2007). Anomalous information reception by research mediums demonstrated using a novel triple-blind protocol. *Explore: Journal of Science and Healing*, 3(1), 23–27. DOI: 10.1016/j.explore.2006.10.004

Botkin, A. L. (2000). The induction of after-death communications utilizing eye-movement desensitization and reprocessing: A new discovery. *Journal of Near-Death Studies*, *18*, 181–209.

Conant, R. D. (1996). Memories of the death and life of a spouse: The role of images and *sense of presence* in grief. In Klass, D., Silverman, P. R., & Nickman, S. L. (Eds.), *Continuing bonds: New understandings of grief* (pp. 179–196). Washington, DC: Taylor & Francis.

Daggett, L. M. (2005). Continued encounters: The experience of after-death communication. *Journal of Holistic Nursing*, *23*, 191–207.

Dannenbaum, S. M., & Kinnier, R. T. (2009). Imaginal relationships with the dead: Applications for psychotherapy. *Journal of Humanistic Psychology*, *49*, 100–113.

Delorme, A., Beischel, J., Michel, L., Boccuzzi, M., Radin, D., & Mills, P.J. (2013). Electrocortical activity associated with subjective communication with the deceased. *Frontiers in psychology*, 4(Article 834), 1–10. DOI: 10.3.389/fpsycg.2013.00834.

Drewry, M. D. J. (2003). Purported after-death communication and its role in the recovery of bereaved individuals: A phenomenological study. *Proceedings of the Annual Conference of the Academy of Religion and Psychical Research*, 74–87.

Hageman, J.J., Peres, J.F.P., Moreira-Almeida, A., Caixeta, L.I., Wickramasekera II, I., & Krippner, S. (2010). The neurobiology of trance and mediumship in Brazil. In S. Krippner and H. Friedman (Eds.), *Mysterious minds: The neurobiology of psychics, mediums and extraordinary people* (pp. 85–111). Santa Barbara, CA: Praeger.

Haraldsson, E. (1988-89). Survey of claimed encounters with the dead. *Omega: Journal of Death and Dying, 19*, 103–113.

Hastings, A., Ferguson, E., Hutton, M., Goldman, A., Braud, W., Greene, E., et al. (2002). Psychomanteum research: Experiences and effects on bereavement. *Omega: Journal of Death and Dying, 45*, 211–228.

Houck, J. A. (2005). The universal, multiple, and exclusive experiences of after-death communication. *Journal of Near-Death Studies, 24*, 117–127.

Kelly, E.W., & Arcangel, D. (2011). An investigation of mediums who claim to give information about deceased persons. *Journal of Nervous and Mental Disease, 199*, 11–17.

Klass, D., Silverman, P. R., & Nickman, S. L. (1996). *Continuing bonds: New understandings of grief.* Washington, DC: Taylor & Francis.

Klugman, C. M. (2006). Dead men talking: Evidence of post death contact and continuing bonds. *Omega: Journal of Death and Dying, 53*, 249–262.

Krippner, S. (2006). Getting through the grief: After-death communication experiences and their effects on experients. In Storm, L. & Thalbourne, M. A. (Eds.), *The survival of human consciousness: Essays on the possibility of life after death* (pp. 174–193). Jefferson, NC: MacFarland & Company, Inc.

LaGrand, L. E. (2005). The nature and therapeutic implications of the extraordinary experiences of the bereaved. *Journal of Near-Death Studies, 24*, 3–20.

Newport, F. & Strausberg, M. (2001). *Americans Belief in Psychic and Paranormal Phenomena is up Over Last Decade: Belief in psychic healing and extrasensory perception top the list.* Retrieved on October 6, 2009 from http://www.gallup.com/poll/4483/americans-belief-psychic-paranormal-phenomena-over-last-decade.aspx

Normand, C. L., Silverman, P. R., & Nickman, S. L. (1996). Bereaved children's changing relationships with the deceased. In Klass, D., Silverman, P. R., & Nickman, S. L. (Eds.), *Continuing bonds: New understandings of grief* (pp. 87–111). Washington, DC: Taylor & Francis.

Nowatzki, N. R. & Grant Kalischuk, R. (2009). Post-death encounters: Grieving, mourning, and healing. *Omega: Journal of Death and Dying, 59,* 91–111.

Parker, J. S. (2005). Extraordinary experiences of the bereaved and adaptive outcomes of grief. *Omega: Journal of Death and Dying, 51,* 257–283.

Roxburgh, E. C., & Roe, C.A. (2011). A survey of dissociation, boundary-thinness, and psychological wellbeing in Spiritualist mental mediumship. *Journal of Parapsychology, 75*(2), 279–299.

Roy, A.E. & Robertson, T.J. (2004). Results of the application of the Robertson-Roy protocol to a series of experiments with mediums and participants. Journal of the Society for Psychical Research, 68, 18–34.

Sanger, M. (2009). When clients sense the presence of loved ones who have died. *Omega: Journal of Death and Dying, 59,* 69–89.

Sormanti, M. & August, J. (1997). Parental bereavement: Spiritual connections with deceased children. *American Journal of Orthopsychiatry, 67,* 460–469.

Julie Beischel, PhD

Certified Mediums' Experiences with the Afterlife

by Julie Beischel, PhD,
with Windbridge Certified Research Mediums
Samara Anjelae, Dave Campbell, Doreen Molloy,
and Stephanie Ann Stevens

from *Winds of Change*, Spring 2010,
Vol. 3, No. 1, pp. 10-11

[Note: Since the original publication of this article in 2010, Stephanie Ann Stevens has retired and we have also certified 14 additional mediums.]

With the release of *Handbook to the Afterlife* by Pamela Heath and Jon Klimo, I thought this would be a good time to get some feedback from the Windbridge Certified Research Mediums (WCRMs) about their views regarding the "Other Side." The current six WCRMs have close to 80 years of combined experience reporting verifiable information about the deceased. Though it is not possible to empirically validate any reports about the afterlife (until we get there ourselves), views based on decades of experience communicating with its inhabitants are—at the very least—quite fascinating. So I asked the WCRMs to send me their answers to the question "what is the afterlife like?"

Dave Campbell pointed out that existence in the afterlife is "very diverse... just as it is on Earth." Dave went on to say, " I find it to be like asking 50 people of different cultures, countries, and belief systems what it is like living on Earth: you would get 50 different perspectives." It appears that answering the question "what is the afterlife like?" may be as difficult as responding to the query "what is life like?" That being said, I did find some common themes among the WCRMs' descriptions.

Both Dave and Samara Anjelae emphasized the parallels of the afterlife and the physical life. Dave called it "very similar to Earth" and Samara described it as "simply 'as above, as below.'"

As below—and as Heath and Klimo report—it seems the afterlife consists of growth and maturation on the part of the individuals "living" there. Dave states, "I am often told that discarnates are in school, learning, and expanding their awareness." Samara also listed stages of evolution as a common theme. Dave went on to say, "I am often told that people who committed suicide are getting counseling, guidance, and healing much like they would if they were here."

Another similarity with an Earthly existence is participating in activities that are enjoyable. Dave says discarnates often show him "things they are doing, such as playing golf" that are "usually what they enjoyed while alive." Samara reports "discarnates describe themselves as busy" and doing activities they were passionate about.

Being with other people is another common theme among the WCRMs' responses. "As I connect with the energy of a discarnate," says Doreen Molloy, "I'm almost always aware of other energies that are also connected in some manner." Stephanie Ann Stevens reiterated this sense of community or family in the afterlife.

There was one common theme that differentiated an afterlife existence to this physical one: the light. Doreen refers to the other side as "a world of light" and, in her list of descriptors, Samara chose "bright" as the first. Stephanie said a discarnate likened the light to the brightness of "a million candles."

The issue of "where" the afterlife is was also brought up. "I am clearly aware that the afterlife is not a place" said Doreen. Her sense is that it exists along with ours. She reports that the deceased "are actually right here among us."

Overall, Samara describes the other side as "beautiful, joyous, and peaceful." And Dave reports that "we can create our surroundings." He says the best example he can provide is envisioning "you are in the safest, most perfect, beautiful place in nature you can imagine, and build a house to your specifics with your imagination, as magnificent as possible, and that is what it is like on the other side. It is filled with non-judgment and unconditional love."

Sounds like a wonderful place.

Julie Beischel, PhD

Recommended Reading: *The Truth About Grief*

from *Winds of Change*, Spring 2011,
Vol. 4, No. 1, pp. 9-10

I recently finished a fascinating new book on a topic around which one our main research programs is focused: grief. This book—*The Truth About Grief: The Myth of its Five Stages and the New Science of Loss* by Ruth Davis Konigsberg—is a quick read that really opened my eyes to numerous aspects of this topic including the bereavement and grief counseling "industries," grief memoirs, the grief culture, grief commercialism, how members of other cultures grieve, and grief related to the events of 9/11/01. The author states, "With this book I hope to offer you a means of escape from our habitual ways of thinking about grief" (p. 197).

Some facts about grief covered in the book that you may not know:

> Elisabeth Kübler-Ross' five stages of grief (denial, anger, bargaining, depression, and acceptance) originated in 1969 with the publication of her book *On Death and Dying*

and proposed that terminal patients facing their own deaths experience each stage in order.

The stages were quickly adopted by the popular media, academia, and society in general and used to describe the experiences of people grieving the loss of a loved one (and later, the loss of just about anything). The pervasiveness of the five stages became readily apparent to me as I read this book when I independently ran across a 2D shooter video game called "Solace" that contains five levels of play, each based on one of the "Five Stages of Grief, as described by Elisabeth Kübler-Ross." The developers report that the theme arose "after one of our team members lost a brother during the game's development."

Research has demonstrated that the stage model of grief does not reflect the reality of most people's experiences. Konigsberg states, "[I]t's more a grab bag of symptoms that come and go and, eventually, simply lift" (p.11).

Most conventions regarding which types and manners of grieving are normal, healthy, or allowed are just social constructs and not based on research or even the majority of cases.

A number of alternative stages (sometimes erroneously listed as Kübler-Ross') have been proposed by other authors and they, too, arose from anecdotes and personal experiences, not research.

In reality, the grieving can go through many different emotions within any given day.

The stage theories seem to have stuck around because they attempt to make sense of something unthinkable, they "bring order to grief's contradictions" (p. 73), they provide reassurance that others have gotten through similar situations, and they allow "us to feel as if we are getting back some control" (p. 79).

Extreme cases of grieving or those resulting from sudden, unexpected, or violent deaths are portrayed more often than milder, more common experiences in books, articles, movies, etc. As a result, many people mistakenly believe those experiences represent the norm.

Medicare hospice legislation passed by Congress in 1982 mandates grief support for a minimum of one year after the death of a loved one.

Acute grief requiring counseling (i.e., prolonged, complicated, or traumatic grief) only affects 10-15% of the bereaved population. Research demonstrates that "...most people are resilient enough to get through loss and reach an acceptable level of adjustment on their own" (p. 198).

A meta-analysis of 61 studies published in 2008 demonstrated that all grieving participants improved regardless of whether or not they received grief treatments or interventions. A similar review published in 2004 made comparable conclusions.

There are too many variables involved to categorize any one type of loss as more difficult than another.

According to researcher George Bonanno, the phenomenon of resilience—reaching an acceptable adjustment to a death within a relatively short time period—"is not rare but relatively common, does not appear to indicate pathology but rather healthy adjustment, and does not lead to delayed grief reactions" (p. 156). Bonanno also determined that laughing, smiling, and repressing negative emotions are more helpful than anger and tears in adjusting to loss.

In closing, I highly recommend this book to those far and wide because even if it does not help you better understand your own grief experiences, it may at least help you in comforting the grieving people you may encounter in your life.

Reference
Konigsberg, R. D. (2011). *The truth about grief: The myth of its five stages and the new science of loss*. Simon and Schuster.

INVESTIGATING MEDIUMS

Julie Beischel, PhD

Survival of Consciousness: Animal Discarnates

from *Winds of Change*, Fall 2011,
Vol. 4, No. 3, pp. 7-13

As our research provides more and more evidence for the survival of human consciousness after physical death (e.g. Beischel, Biuso, Boccuzzi, & Rock, 2011, and Beischel & Rock, 2009), one of the next logical questions involves the potential survival of non-human animal consciousness. If non-locality is simply a characteristic of consciousness (a theory supported by various aspects of psi research), then logic dictates that any and all consciousnesses can potentially exist without a physical body. Below I will explore the biological, evolutionary, and philosophical support for this theory as well as report on the experiences of several Windbridge Certified Research Mediums (WCRMs) with discarnate animals. In addition, comments from a bereavement expert will shed light on grief after the death of a companion animal.

Do Animals Have Consciousness?

Stephen Jay Gould, paleontologist, evolutionary biologist, and historian of science, once stated, "The most important scientific revolutions all include, as their only common feature, the dethronement of human arrogance from one pedestal after another of previous convictions about our centrality in the cosmos." Numerous other authors, philosophers, and researchers have also shined light on the fallacy of human "centrality" or superiority over other species and found this "speciesism... analogous to racism and sexism and therefore unacceptable" (Sharpe, 2005, p. 7). It appears that—at the very least, certain—animals are quite similar to humans in their perceptions, emotions, communication, logic, relationships, individuality, and psychology.

For example, it has been argued that animals do not have emotions and simply react instinctually to external stimuli. However, though animals do not have the linguistic capabilities to label their emotions as "happy," "sad," or "angry," their bodies experience those emotions just like human bodies do. Behavioral neurobiology expert Antonio Damasio and his co-researchers have demonstrated through work with individuals who have sustained damage to portions of the prefrontal cortex that emotions begin in the body and then are recognized by the brain. That is, rather than me thinking, "I'm scared" or "That sure is a large gun" and my body reacting by releasing adrenaline; my body first reacts to the fear-inducing stimulus by releasing adrenaline and afterward my brain labels it as fear. Emotions happen in the body. The brain might then assign words to them, but they happen in the body. So even without a pre-frontal cortex and a labeling system, animals still experience emotions where they happen: in their bodies.

In a recent Big Think interview, Damasio stated:

> ...your emotion of joy and mine are going to be extremely similar... And it's going to be the same across even other species. We may smile and the dog may wag the tail, but in essence, we have a set program and those programs are similar across individuals in the species. Then the feeling is actually a portrayal of what is going on in the organs when you are having an emotion.

I have second-hand experience with how this system can go awry. My mother did not have pre-frontal cortex damage, but the disorder that affected her—alexithymia—causes an alteration in the connection between the body's experience of an emotion and the brain's recognition of it. This was probably best demonstrated in a clinical video of my mother talking with a psychiatrist in which he stops her to ask, "Do you know that you are crying?" She was, in fact, completely unaware that she was. Needless to say, the knowledge that she was sad was even further from her awareness.

Her body was experiencing sadness, but she did not have the ability to cognitively accept what that meant. I'm sure you can see the fallacy in saying that zebras, dogs, cows, finches, and my mother do not have emotions because they cannot cognitively define or label them.

Animal Discarnates

If we are not so different from our animal counterparts, then the question arises as to their presence with us in the afterlife. In his comprehensive tome *Is There an Afterlife?*, David Fontana writes:

> It is a sign of our arrogance as a species that we assume animals have no self-consciousness, no awareness of their own mortality, and no profound emotions and feelings comparable to our own. All too often this arrogance leads us all to suppose that if survival is a fact, it applies exclusively to humans (p. 11).

Indeed, there are numerous cases arising from, for example, mediums, channeled information, instrumental transcommunication (ITC), and near-death experiences (NDEs) regarding the presence of animals in the afterlife. In *The Handbook of Near-Death Experiences*, Sutherland (2009) provides this example from a child's near-death experience:

> In [an] interesting manifestation of divine guidance and comfort, seven-year-old Pat was greeted not by an "angel" or a "beautiful lady" or even by friendly "light figures" but by two much-loved family pets who had died four years prior to his near-drowning (p. 98).

In *Handbook to the Afterlife*, Heath and Klimo (2010) report that medium George Anderson has:

> channeled a number of souls that speak of being greeted by animals. In one case the first spirit to greet the soul was of a beloved family pet. Knowing the animal had already died made it clear to the spirit what was going on, and he happily followed it into the Light. In another case, a woman was greeted by a dog, which was a welcoming and comforting sight for her. Anderson feels that animals may be especially frequent greeters to suicides because of their ability to be nonjudgmental and give unconditional love. This may be true even for those who have never had a pet. They may still feel like following a puppy or kitten to leave the vibrational region near the earth plane and complete their transitions to the afterlife (pp. 99-100).

WCRM Doreen Molloy devotes Chapter 7 of her book *Proof Positive: Metaphysical Wisdom*, to the topic "Consciousness in Other Life Forms: Do animals have souls?" and states:

> If we are able to communicate with people who have crossed into the spirit realm via a "mind to mind" connection, theoretically, why couldn't we be able to do that with an animal? The verbal language that we're accustomed to using is not engaged when we are receiving a message from a human spirit... I believe that communication with animals can take place, utilizing the same principles, whether they are alive or not. (p. 141-142)

Similarly, in his book, *Bridge to the Afterlife: A Medium's Message of Hope & Healing*, WCRM Troy Parkinson states:

> ...all animals have a spirit and, just like humans, when an animal's physical body dies, their spirit body lives on... Whether it's dogs, cats, or fish, spirit is spirit, love is love, and connections are made between two worlds (137-38).

Along these lines, Fontana (2005) reports that "mediums sometimes claim to see or hear dogs, cats, and horses in the afterlife, and the descriptions given of these animals fit the deceased pets of sitters" (p. 456). We have similarly noted the presence of and accurate descriptions of these animals—as well as birds and rodents—in our own research with mediums.

None of this evidence was surprising to me. During the first (and only) official mediumship reading I had (and which started me on this research path), my mother was accompanied by two dogs: our family Dalmatian and the small, beige dog who we used to dog-sit regularly, each of who followed my mom around like happy, furry shadows. (All of that second dog's people are still living, so I guess the dog-sitting continues on the other side.)

Recently, I asked the Windbridge Certified Research Mediums (WCRMs) about their experiences with discarnate animals.

WCRMs' Readings

When asked how often companion animals spontaneously appear in readings for sitters, the WCMRs reported animals in an average of more than 30% of their readings. Daria Justyn stated that, "80% of the time animals show up and are usually very excited to do so." Ankhasha Amenti reported that, "Even if the purpose of the reading is not mediumship or spirit contact, they seem to pop in." She added, "Since I own a bird, two cats, and a dog, maybe they feel I am open to their visit." Renata Santorelli, another animal lover, reported that animals appear in her readings 85% of the time and that "the animals usually show up in the readings before a sitter asks. Animals are amazing!"

Eliza Rey said, "Sometimes the animal is not the sitter's animal, but an animal that may have lived on the property where they reside. The animals usually give the sitters a message of love and companionship. At times, the animal comments on the other living animals that either replaced them or are present. A few times, an animal is getting ready to pass and I pick up on their message to the sitter."

The WCRMs reported that sitters requesting that they specifically connect with a deceased companion animal occurred far less frequently and usually during readings for deceased people. This is understandable considering that the WCRMs also reported sitters being surprised when an animal appeared in a reading. Sitters may be unaware that mediums can communicate with deceased animals.

Carrie Cox's experience has been that, "Clients do not typically think to ask to talk with a deceased companion animal in isolation. They are more likely to ask in conjunction with connecting to a deceased human family member. They will ask if the beloved pet is with their loved one."

Nancy Marlowe's comments were similar to Carrie's: "I can't remember anyone asking me to connect with an animal unless another animal has already come in unexpectedly after the reading has started. Most of my clients are joyfully surprised when a pet comes through. Many clients express gratitude to have heard from their pets. They smile or cry." Joanne Gerber noted, "I do get occasional requests. However, like with human discarnates, mediums cannot 'call up' a specific animal discarnate to come through during a session but I will make the attempt to see what comes through. I have found that it is most evidential if we do not know what breed, name, or what type of pet that the sitter wants to connect to, using the same protocol as in human discarnate communication."

Regarding which types of animals communicate, Traci Bray stated, "dogs, cats, ponies, rabbits, snakes, goldfish—the gamut have come through both independently and when summoned." Joanne Gerber added, "I've experienced communication with all types, including hermit crabs and tarantulas!" Other WCRMs mentioned monkeys, pigs, birds, and horses in addition to the more common dogs and cats.

I also asked the WCRMs to share specific stories about their experiences with animal discarnates and received many fascinating responses. Below are a few of their comments in no particular order.

Joanne Gerber: Just recently, I had a session with a dog that had passed. The dog's message to the sitter was, "Please don't be angry with me. I'm sorry about the paintbrushes." The sitter, an artist, validated that her dog had chewed up one of her new paintbrushes before he passed.

Marissa Ryan: A client called about her mother who had passed and a monkey came in to communicate. It described foods it ate, how it died, and each of its toys. She then confirmed her mother did have a pet monkey. I was blown away.

Laura Lynne Jackson: There was a reading I did recently where the client's mother showed me a little white dog with her and the specific message was that the dog had been there to meet and cross her when her time to go to the other side came, making the transition smooth and loving. The client was greatly comforted by that as she said her mother "lived for that little dog."

Stephanie Ann Stevens: I was scheduling a reading on the phone and smelled peanut butter over and over and finally asked the client if she had the nickname of 'Peanut.' She exploded with joy as her little dog, Peanut, had passed a month before.

Renata Santorelli: Immediately during the start of a reading, I kept being shown a big beautiful black stallion! I shared with the sitter what I was seeing and that "I'm being told to say 'Mama, it's Mister Prince!'" The sitter was shocked and beside herself with tears. I then shared that the horse wanted me to say, "Thanks for saving me and thanks for the biscuits!" The sitter knew exactly what that meant. He shared with me that she had saved him from being slaughtered and that he loved her homemade biscuits. She was more than overjoyed and I was amazed at how clearly I was seeing and hearing him.

Tracy Lee Nash: During a reading, a son had come through and was making a reference about "Peter" which was followed by a squawk such as what a parrot would make. The son wanted his mom to know that he and the bird were fine, as she took care of the bird after her son had passed away until the bird passed as well. She was hoping they were together as her son adored the bird. She stated after the reading that she would hear Peter's squawking throughout the house and it was a validation for her that he was all right.

Grieving Those Animals Who Have Passed

In *Proof Positive: Metaphysical Wisdom*, WCRM Doreen Molloy reports:

> The grieving process for losing pets is not much different than it is for people; the bonds are sometimes just as strong as experiencing a human loss and if that's the case, we need to know that they are all right, too (p. 136).

WCRM Daria Justyn concurs: "An opportunity to reconnect with a departed pet has such a healing effect on people and instills a sense of comfort about our loved ones on the other side."

The Encyclopedia of Death & Dying reports:

> Because many people form deep and significant emotional attachments to their pets, at any

given time the number of people suffering from grief in relation to the loss of a pet is quite high. Pet loss has been shown to potentially have a serious impact on an owner's physical and emotional wellbeing.

This source goes on to state:

Despite the fact that the resolution of the grief often surpasses the length of time seen with human losses, the easy accessibility and replacement of the lost animal often provokes hidden grief reactions. Grief may also be hidden because of the owner's reluctance and shame over feeling so intensely over a nonhuman attachment. People who have lost a pet may repress their feelings, rationalize or minimize their loss, or use denial as a way to cope.

I recently spoke with Lavon Switzer, LCSW, who has worked with the bereaved for 39 years and recognizes the need for specific support for individuals mourning the loss of a companion animal. As an expert in grief, Lavon is often contacted by other therapists and counselors in private practice asking where they can refer someone who is mourning an animal and found that there were almost no resources to offer them.

Lavon notes that the experience of losing an animal is comparable to losing any other family member.

Grieving an animal may, at times, be more difficult because our society does not recognize how similar the loss is to losing a person. This makes it difficult for some people to understand the grieving process of individuals who have lost "just" a dog or "just" a cat.

Lavon reports, "I don't work with somebody who has lost an animal any differently than I work with somebody who has lost a family member because it's the same. I help people learn what to expect from being in the state of grief and to understand that they're not going crazy." She also stressed that she regularly fields questions from people who think there is something wrong with them because they haven't gone through what they erroneously believe are the stages of grief and she must reassure them that what they are experiencing is normal. (For further reading on this issue, see *The Truth About Grief: The Myth of Its Five Stages and the New Science of Loss* by Ruth Davis Konigsberg and/or my review of it.)

Lavon noted that "even though we all grieve differently, a majority of us experience symptoms like forgetfulness, the inability to focus, sleep and eating disturbances, and experiences of the person or animal after the death."

For further insights on coping with the grief after losing a companion animal and "with the difficult decisions one faces upon the loss of a pet," Moira Anderson Allen, MEd, of Pet-Loss.net offers Ten

Tips on Coping with Pet Loss and answers such questions as 'Am I crazy to hurt so much?' and 'Will my other pets grieve?' Grief expert and author Lou LaGrand also offers online advice on how to assist others with animal losses in his article, "How to Help Someone Mourning the Death of a Pet."

The Windbridge Institute bookstore offers several books on the topic of grief after the loss of a companion animal that may serve as comforting gifts for those in mourning.

Visit Anthrozoology.org for listings of academic research papers and abstracts dealing with pet loss, grief, euthanasia, and thanatology. "Anthrozoology is the study of the relationships between humans and animals. It is unique in that it studies the role of animals in the lives of humans, and vice versa."

Spontaneous After-death Communications (ADCs) with Animals

During our research, we have heard numerous anecdotes about people feeling the presence of deceased companion animals or having other after-death experiences related to animals.

WCRM Ankhasha Amenti had her own spontaneous experience with a discarnate cat:

I was the only guest at one of my favorite haunted bed and breakfast homes, Thayer's in Annandale, Minnesota, and my room was locked. Around 4 a.m., I woke up feeling there was something in my room, and I turned on the light. At the foot of my bed appeared little paw prints (visually, not a psychic impression I had) and just like a little kitty would, they climbed up towards me, snuggled up on my shoulder, and started to purr really loudly. I started to laugh out loud because it was so fun for me even as a psychic to see the little paw prints, and hear the purring, not just sense them. When I told the owner in the morning, she said it was most likely the ghost of one of her beloved cats that frequented that room and was always climbing on the bed and snuggling with people. She told me he would make appearances there when he felt the guest would allow him to sleep with them. It really was an amazing experience, because it was like a "special effect" from the other side!

WCRM Daria Justyn also had a personal experience in which her deceased cat, Shadow, appeared to her. She commented, "I absolutely love the thought that the animals that were entrusted to us here on Earth are happy and whole and still feeling our love on the other side."

In his article "Animals and Ghosts: The incredible connection between animals and the spirit world," Stephen Wagner notes:

> The ghosts of animals may be as common as the ghosts of humans. There are many reports from people who have sensed, felt, smelled, heard and even seen the spirits of recently departed pets... Like human ghosts, animal ghosts can sometimes appear to lend comfort.

In her book, *When Spirits Come Calling: The Open-Minded Skeptic's Guide to After-Death Contacts*, Sylvia Hart Wright (2002) states that "several of my 78 interviewees reported deeply emotional contact experiences with precious, departed dogs" (p. 155). In the case described below, the interaction was—quite literally—life-saving.

Bob was a soldier fighting in the jungles of Vietnam. When several of his friends were killed after their unit was hit by enemy fire, the blood from a head wound Bob had sustained made it difficult for him to see and he became disoriented. It was then that Bob's childhood dog, Rusty, appeared to him barking and leading him down a path toward help. Upon returning home, Bob was told by his family that Rusty had died suddenly of a heart attack in Texas at the same date and time that he had appeared in the jungle in Vietnam. "And Bob always will believe that his dog died of a heart attack when he was wounded... and then went to save him" (p. 158).

Few researchers have investigated these phenomena specifically, but researcher and Windbridge Institute Scientific Advisory Board Member Erlendur Haraldsson collected data in 2006 from 991 people living in Iceland regarding their psychic experiences. The survey included the question "Have you ever been in the presence of a deceased animal?" to which 9% of respondents reported that they had (6% of men, 12% of women). Haraldsson (2011) reports that "of these instances, 87% concerned pets. Mostly they were seen (39%), heard (39%), or a touch was felt (31%)" (p. 81).

Windbridge Institute Research

A new interdisciplinary field called Human-Animal Studies (HAS), "one of the most rapidly growing fields of intellectual inquiry today" (DeMello, 2010, p. xii), takes the human-animal relationship as its central focus. Scholars in disciplines as diverse as anthropology, art history, drama, philosophy, social work, and veterinary medicine all approach this subject from very different perspectives and with different methodologies, but all are interested in analyzing the complexities of the human-animal relationship (DeMello, 2010, p. xi).

The addition of HAS into parapsychology in general and survival research specifically is a logical expansion. Although the presence of psi in human relationships with living animals has been studied for numerous decades (reviewed in Dutton

& Williams, 2009), studying those relationships once a companion animal has died is a worthy pursuit.

Individuals with close relationships to animals experience them as family members and grieve their losses similarly to the loss of human spouses, parents, children, etc. It follows logically that the motivation to communicate once an animal has died—from both sides of the veil—would be similar to that which arises after the death of a human loved one.

Earlier this year, I received a grant from the Survival Research Committee of the Society for Psychical Research (SPR) for a project in which we will use the standard Windbridge Institute quintuple-blind mediumship reading protocol to investigate WCRMs' abilities to report accurate and specific information about deceased companion animals.

I hope you have found this discussion of animal consciousness, animal discarnates, the experiences of WCRMs and other individuals, and grief surrounding the loss of an animal interesting. Thank you for supporting research at the Windbridge Institute.

References

Beischel, J., Biuso, M., Boccuzzi, M., & Rock, A. (2011, June). *Anomalous information reception by research mediums under quintuple-blind conditions: Can the mind exist without the body?* 30th Annual Meeting of the Society for Scientific Exploration, Boulder, CO.

Beischel, J., & Rock, A. J. (2009). Addressing the survival vs. psi debate through process-focused mediumship research. *Journal of Parapsychology, 73,* 71–90.

DeMello, M. (Ed.) (2010). *Teaching the animal: Human-animal studies across disciplines.* New York: Lantern Books.

Dutton, D., & Williams, C. (2009). Clever beasts and faithful pets: A critical review of animal psi research. *Journal of Parapsychology, 73,* 43–68.

Fontana, D. (2005). *Is there an afterlife? A comprehensive overview of the evidence.* Blue Ridge Summit, PA: NBN.

Haraldsson, E. Psychic experiences a third of a century apart: Two representative surveys in Iceland with an international comparison. *Journal of the Society for Psychical Research,* 75.2, 76–90.

Heath, P., & Klimo, J. (2010). *Handbook to the afterlife*. Berkeley, CA: North Atlantic Books.

Konigsberg, R. D. (2011). *The truth about grief: The myth of its five stages and the new science of loss*. New York: Simon & Schuster.

Molloy, D. (2004). *Proof positive: Metaphysical wisdom*. Bloomington, IN: Authorhouse.

Parkinson, T. (2009). *Bridge to the afterlife: A medium's message of hope & healing*. Woodbury, MN: Llewellyn Publications.

Sharpe, L. (2005). *Creatures like us?* Charlottesville, VA: Imprint Academic.

Sutherland, C. (2009). "Trailing clouds of glory": The near-death experiences of western children and teens. In Holden, J. M., Greyson, B., & James, D. (Eds.). *The handbook of near-death experiences: Thirty years of investigation* (pp. 87–107). Santa Barbara, CA: ABC-CLIO.

Wright, S. H. (2002). *When spirits come calling: The open-minded skeptic's guide to after-death contacts*. Nevada City, CA: Blue Dolphin Publishing, Inc.

INVESTIGATING MEDIUMS

Julie Beischel, PhD

Among Mediums

A Scientist's Quest for Answers

Julie Beischel, PhD

Julie Beischel, PhD

Praise for *Among Mediums...*

"Dr. Julie Beischel is a courageous, innovative pioneer who has brought immense respectability to a controversial area. If you think modern science has proved beyond doubt the finality of physical death, think again. Dr. Beischel's research... points like an arrow to an aspect of consciousness that survives. Her contribution is huge, because throughout human history the terror of death's finality has caused more suffering than all the physical diseases combined... In the telling, Beischel has established a new genre: science-and-humor. Who said writing about dead people has to be morose and gruesome? ...Be prepared to laugh. This, dear reader, is science writing at its best."
–Larry Dossey, MD, author of *One Mind: How Our Individual Mind Is Part of a Greater Consciousness and Why It Matters*

"Julie Beischel and her team at the Windbridge Institute have joined a unique group of scientists over the past century who have studied mediums... Beischel's team has taken on the challenge in a serious and rigorous way, and the results of their efforts are intriguing... Easy to read with Beischel's sense of humor liberally sprinkled throughout the text, making it both authoritative and accessible."
–Dean Radin, author of *The Conscious Universe: The Scientific Truth of Psychic Phenomena*

"Finally, a book about mediums that explains real science and does not require matching t-shirts or a clever acronym! Smart girls ROCK!!"
–Katie Mullaly, co-author with J. Patrick Ohlde of *Scare-izona* and *Paranormal Pandemic*

"The book is a little gem - short and succinct, highly readable, and packed with interesting insights... Sceptics passionately believe in the efficacy of JREF's Million Dollar Challenge as a statement about psychic ability, but this is truly and transparently scientific in a way that the Challenge cannot remotely claim to be. Beischel is to be congratulated for her dedication to understanding a type of human experience that, for largely ideological reasons, science as a whole has shunned."
—Robert McLuhan, Author of *Randi's Prize*

"...entertaining, humorous, unpretentious, almost breezy... The meatiest part of the book is Beischel's discussion of the protocols employed by the Windbridge Institute to rule out information leakage in testing mediums... The precautions are almost paranoid in their elaborateness, and it's hard to see how even the most determined skeptics could poke holes in the procedure (though I'm sure they will try)... *Among Mediums* is an excellent introduction to scientific research into purported communications from 'the other side' and one that should appeal to intelligent, open-minded readers of all backgrounds."
—Michael Prescott, Author of *Grave of Angels*

"Very readable, well-researched, and with a delightfully dark sense of humor... While not proving or disproving the afterlife, Beischel presents a reasonable case for a phenomenon that we don't currently understand but could have some profound social applications if taken more seriously by the scientific community."
—Ryan Hurd, author of *Big Dreams: Psi, Lucia Dreaming and Borderlands of Consciousness*

"Easy to read and understand and with a light-hearted touch, Beischel's book is a wonderful introduction to the application of science to a human experience that is thousands of years old yet which has been ignored or dismissed by most of the scientific community. I highly recommend this book for anyone interested in mediumship, in the mounting evidence for life after death, in the very real potential for mediums in the grieving process, and in how science can be applied to the so-called paranormal."
–Prof. Loyd Auerbach , MS, Atlantic University & JFK University

"This is the finest and clearest piece you will find on this subject matter as it illuminates with humor and expertise the thorniest issues everybody is afraid to ask about survival. Probably no one has done as much detailed scientific research as Dr. Beischel in such an original way. Her results are unquestionably authentic and her unique direction unparalleled. I could not recommend this read and this research more unreservedly. With my many years of work in the arena of grief healing, I can tell you Beischel's work will transform our understanding of the freedom from fear that is available to all of us."
–Samadhi Longo-Disse, M.Div, STM, PhD, Grief Counseling Professional

"A short and concise but absorbing book... It has been only in the last 15 or so years that a few courageous scientists have risked showing any interest in studying mediumship and picking up where those esteemed researchers of yesteryear left off. Dr. Beischel is one of those few."
–Michael E. Tymn, author of *The Afterlife Revealed: What Happens After We Die*

"Dr. Beischel takes the scientific testing of mediums to a new level and presents her findings in this easy-to-read, fact-filled and entertaining book. She answers many of the questions that have gone unanswered for far too long, and removes any reasonable doubt about the validity of these communications."
–Bill Kaspari, author of *The Galilean Pendulum: A New Science Reveals An Unseen World*

"Dr. Beischel has managed to unlock many of the mysteries surrounding mediumship and explain the common misconceptions about the process. Perhaps even more important than the scientific research are her queries about its implications and how the knowledge can be used for the betterment of humanity."
–Robert Ginsberg, Vice-president, Forever Family Foundation

The Windbridge Institute, LLC
1517 N. Wilmot Rd. #254
Tucson, AZ 85712
http://www.windbridge.org/
info@windbridge.org

Text copyright © 2013 by Julie Beischel, PhD
All Rights Reserved

First e-book edition January 2013

The names of some individuals and locations in this book have been changed. The information in this publication is provided "as is" without warranty of any kind, either express or implied, including but not limited to the implied warranties of merchantability and fitness for a particular purpose. Under no circumstances shall the Windbridge Institute, LLC, nor any party involved in creating, producing, or delivering this publication be liable for any damages whatsoever including direct, indirect, incidental, consequential, loss of business profits, or special damages. Always consult with your physician or other qualified healthcare provider when seeking treatment options.

*To mediums everywhere
who use their natural abilities
to help people
even when faced with slander,
derision, and ignorance:
Thank you.*

Julie Beischel, PhD

Among Mediums Contents

- ∞69∞ Chapter 1: On the Same Page
- ∞79∞ Chapter 2: How Did I Get Here?
- ∞105∞ Chapter 3: Science for All
- ∞119∞ Chapter 4: Information Research Program
- ∞133∞ Chapter 5: Operation Research Program
- ∞145∞ Chapter 6: Application Research Program
- ∞161∞ Chapter 7: So What?
- ∞166∞ Appendix: Cited Resources and Materials

Julie Beischel, PhD

❧ 1 ❧

On the Same Page

Why is it so noble and respectable
to find whence man came,
and so suspicious and dishonorable
to ask and ascertain whither he goes?
—James Hyslop

For the last 10 years, I have been studying mediums, people who experience hearing from the dead. I have been examining the information mediums report as well as their unique experiences and the socially relevant purposes their readings might have. Over the years, I have encountered many misconceptions about mediums and about mediumship research from the media, the scientific community, and the general public. With this book, I hope to dispel some of those myths.

The first question I am usually asked when people find out what I do is: "How did you get started in all this?" The second question is: "What have you found in your research?" This book discusses the answers to those two questions.

Before we get to those discussions, however, there are some things that I need to disclose to you:

I have attempted to write this book like I was sharing this information with you over coffee (though I hate coffee, so I'll be having lemonade and a cookie) and not like I talk when I present my findings at academic meetings. And because we're both busy and we can't sit in the coffee shop indefinitely, I have tried to be succinct. (I like to hear myself talk—or read myself writing—as much as the next academic, but I tried to rein it in for this book.) Where applicable, I have pointed you to other resources for further reading on specific topics (there is a complete list of the cited books, articles, and other materials at the end), but all the necessary and sufficient pieces of the story have been included here.

I think there is too much horror in the world to be serious all the time, so I tend to joke around quite a bit. I take my research about consciousness, death, and grief seriously, but I don't think a solemn attitude is required to do so. As George Bernard Shaw once said, "Life does not cease to be funny when people die any more than it ceases to be serious when people laugh." I hope you will not find my humor irreverent.

With topics like mediumship, grief, and life after death, it is easy to get caught up in the drama. I have made every attempt to limit my use of pretentious, sensationalist, theatrical, and overly evocative language in this book. I'm sure you are aware that there are plenty of other books that fulfill those needs.

This is not a book for academics (though most of them are people as well, so they, too, may enjoy it and even learn something). This is a book for people who are interested in what science has to say about modern mediums. It's for people who have seen what television producers imagine appeals to the public and who now want the real story. However, like I said above, links to the original scientific publications are included throughout for people who would like further details.

This is also not a book defending parapsychology as a valid and rigorous form of scientific inquiry. There are plenty of other authors and researchers who have taken up that fight. (If you are interested in this topic, I suggest you check out *Entangled Minds: Extrasensory Experiences in a Quantum Reality* by Dean Radin, *Randi's Prize: What Sceptics say about the Paranormal, Why they are Wrong and Why it Matters* by Robert McLuhan, and/or the other materials and authors listed in the appendix.)

This book was not written to change minds. There are few undecided voters when it comes to these topics. There are believers and there are deniers. I am a scientist, so I fall somewhere in the middle. This book simply discusses what my research demonstrates.

Science is just one way of knowing. It (and I) can neither refute the existence nor defend the reality of your experiences or what you know in your heart to be true.

The mediums with whom I work practice what is called "mental" or "clairvoyant" mediumship. During readings, they enter a semi-altered state of consciousness but are, for the most part, fully conscious and aware and can remember the reading afterwards. Unlike during "trance" mediumship, the bodies and voices of mental mediums are not taken over by the communicating entity and they do not require complete darkness. Readings with mental mediums do not (usually) involve the phenomena that have been historically reported during sessions with individuals called "physical" mediums: objects moving around the room; unexplained lights; independent voices, raps, or other sounds; ectoplasm (a vapor-like substance exuded from the body of the medium); and apports (objects that mysteriously appear). Mental mediums have multi-modal experiences (see Chapter 5) of communication from 'the other side' and verbally report those experiences and often their interpretations of symbolic information.

Although I have been performing research with mediums full time for about a decade, there is only one of me (and, as you will learn in Chapter 2, I am not a prime specimen either physically or emotionally). The other researchers on our team are volunteers and/or have other full-time jobs or other research projects of their own. In addition, the resources that support this type of research are limited (mainly, private foundation grants, memberships, and absolutely none of your tax dollars), so I must spend some of my time on tasks optimizing our public visibility (which translates to more memberships and other opportunities for funding) in order to continue purchasing shelter and food. It's not an ideal situation, but I am not complaining; it is what it is. This topic is worthy of investigation and I seem to be good at investigating it. I just want you to be aware of the larger reality under which this research takes place. We are simply not able to do all the studies we would like. So when you think to yourself, "I wonder if they've ever done/asked/thought of [blank]," please know that the answer is most likely, "Yes, we have, but there hasn't been the time or resources to do/answer/address [blank] so we've put it on the list of tasks to tackle in the future when there is more funding to hire more researchers." Rest assured, however, that I have plenty of unique and interesting research findings to share with you here.

The information contained in this book should not to be used as a substitute for medical or psychiatric advice or be used to replace the guidance of appropriate professionals.

Some of the names have been changed to protect the innocent (as well as the guilty).

The existence of life after death has not been proven, but in order to facilitate a smooth read, I have chosen not to preface each use of "communication with the deceased" and similar phrases with "ostensible," "alleged," "supposed," "perceived," or the like.

When I was getting feedback about the initial versions of this book, it was suggested to me that I include more stories about the research participants and their experiences. And while I understand that these types of stories add a personal aspect to the science, those aren't my stories to tell. I can only share with you my data and my experiences.

A final note: You may have noticed my extensive (or excessive) use of parentheses. You may choose to read the parenthetical text as if I were leaning across the table to whisper my snarky, tongue-in-cheek comments so that only you can hear them (though some parentheses are just parentheses), but if you'd prefer to disregard my commentary (as delightful as it is), please skip over the parenthetical bits. I did, however, temporarily purge my extensive (or excessive) use of swear words during the writing of this book. You will not find any F-bombs here (parenthetical or not).

The Survival Basics

With the disclaimers out of the way, let's start with some basic facts. The official scientific term we use instead of "the Afterlife" or "Life After Death" is "the survival of consciousness" or simply "survival." Like with most scientific jargon, this term describes the phenomenon without really committing to anything more. Just as 'unidentified flying object' does not convey the nature of the object or where it came from (only that it flies and is unidentified), 'survival of consciousness' does not commit to definitions of what survives, how it survives, or the nature of its existence.

Survival research involves many different subcategories including children who remember past lives (also called 'cases of the reincarnation type' or CORT); hauntings and apparitions; near-death experiences (NDEs) and out-of-body experiences (OBEs); instrumental transcommunication including electronic voice phenomena (ITC and EVP); deathbed visions; spontaneous cases of after-death communications (ADCs); and mediumship. As you may have gleaned from its title, this book is about mediumship, but for an extensive discussion of all of these topics (what we call 'planks in the survival platform'), you may want to check out *Is There an Afterlife? A Comprehensive Overview of the Evidence* by David Fontana. And for an inside look at the historical mediumship research performed by the Societies for Psychical Research in London and the US in the 1880s, I recommend *Ghost Hunters: William James and the Search for Scientific Proof of Life After Death* by Deborah Blum.

The Lows and Highs of Mediums

One undeniable fact is that there are people who claim to experience regular communication with the deceased. Those people are called mediums. The word 'regular' is important there: one or a few mediumistic experiences do not make a medium (or most of us would be called mediums). Mediums are different from psychics; psychics experience regular access to information about people, the future, or distant locations they couldn't otherwise know. Mediums can also be psychic, but psychics aren't necessarily mediums. Mediumistic experiences are common and normal. It seems to be like any other perception (vision, hearing, etc.); it is possible for many human brains to see, but a lesser number have 20/10 visual acuity and even fewer can make a living based on their unique perceptive abilities.

Also a fact is that there are people who want to hear from their deceased loved ones and visit mediums to do so. During research, those people are called sitters. The process during which a medium reports information to a sitter is called a reading.

If all of the secondary facts related to those initial realities (for example, consciousness continues to exist after death and it can talk to a medium, that medium can hear the consciousness, etc.) were accepted at face value as normal and real (as they are in many cultures around the world), that would be the end of the story.

But in the Western world, phenomena not easily explained by the traditional, established sciences are usually dismissed as impossible. Usually, people who believe in phenomena like mediumship are labeled ignorant, gullible, or delusional, and the unfortunate individuals who experience mediumistic communication are called frauds, con-artists, schizophrenics, evil, or worse.

Now, what if we calmed down, put aside our assumptions about how the world works, and actually applied the scientific method to the phenomenon of mediumship? Well, I did just that, and this book reviews what I discovered.

In Chapter 2, I review how I got into this field and how my experiences convinced me that there was something worth investigating. Chapter 3 explains, in simple terms, how the scientific method can be applied to the phenomenon of mediumship. In Chapter 4, I review the results from studies of the accuracy of mediums' statements. Chapter 5 examines the unique experiences, physiology, and psychology of mediums. The practical social applications of mediumship readings are discussed in Chapter 6. And in Chapter 7, I attempt to answer the question, So What?

Julie Beischel, PhD

❦ 2 ❦

How Did I Get Here?

Every exit is an entry somewhere.
–Tom Stoppard

Of all the chapters in this book, this one was the hardest to write. I struggled with whether or not to even include it. Half of me is quite private and is entirely bewildered by the current aspects of our culture that involve the sharing of personal details all over cyberspace. However, the other half of me believes very strongly that you should always question the source of new or contrary information; that is, knowing the motivations driving the person making a claim is essential before one can accept the claim. That latter half of me thinks it is important for you to know a little bit more about me personally before you can accept the conclusions I have drawn about mediums.

As you might have guessed, that latter half won the tug-of-war and I have included here the information about me that will help you determine if I am a credible, impartial source or if I have some kind of vested interest in the outcome of my research. I hope that you will notice that because of difficult family relationships, it would actually be easier on me if death were simply the end. Unfortunately for me, that final end doesn't seem to be where the data are pointing.

A Traditional Start

I received my undergraduate degree 7,000 feet above sea level. It was a Bachelor of Science (BS) degree in Environmental Sciences with an emphasis in Microbiology from Northern Arizona University in Flagstaff. I remain a militant environmentalist (as well as a vegetarian) today, but once I took Medical Microbiology in college, I knew my passion was more in line with cells, blood, and physiological processes than with rocks, trees, and the water cycle. However, two main environmental principles really stuck with me: 'The solution to pollution is dilution' and, similarly, 'At the root of all environmental problems is overpopulation.' (Thus, to this day, the idea of a decimating plague doesn't sound so awful to me. Also, I only buy 100% recycled toilet paper and tissues because I don't think a tree should die just so that I can wipe my, um, nose.)

With my new found interest in human biology, I looked into going to medical school, but a couple of realizations steered me away from that path. I recognized that (a) they won't teach you all the interesting things about the human body unless you agree to touch one or two (and the bodies you have to touch will be sick or even dead!) and (b) I have a proclivity for research: hypothesizing and experimenting. Medical school, it seemed to me, was far more about memorization and regurgitation than about deduction and logic. I ended up having some academic and personal interactions with medical students and I was not, on average, impressed with their use of critical thinking.

So I started looking into graduate programs. I did not, at that time (or when I had gone to college), have the psychological resources to even imagine being more than a couple hundred miles from my parents in Phoenix, so I kept my search within Arizona. Then, when I was 19, I woke up one morning with only peripheral vision in one eye; I couldn't see what I was looking at, only the things around it; the center of my field of vision was simply gone; not black, not blurry, gone. After a spinal tap and an MRI, I was diagnosed with multiple sclerosis (MS) and instructed to avoid infection and keep my stress level to a minimum. (Graduate school at a hospital sounds like a great place to have that life. Yeah, right. I went anyway.)

I ended up getting my Doctorate of Philosophy (PhD) in Pharmacology and Toxicology with a minor in Microbiology and Immunology/ Immunopathology (I petitioned to have my pathology courses serve as minor credits) from the

University of Arizona (UA) in Tucson. It was a hard-fought victory that took seven years. Two different times, I was unable to use months of data due to facility contamination, my original PI (principal investigator; the person whose lab I worked in and who oversaw my research) took a leave of absence after a sexual harassment scandal, my MS acted up, and my mother committed suicide.

Touched by Death

There's no way to ease gently into that last item, so I figured I'd just spring it on you. The situation was, as you would expect, difficult. Happy, psychologically healthy lives rarely end that way, so you would be accurate in concluding that my relationship with my mother had been strained. Thus, her death was sadly a relief. It was a horrifying shock, but the healing, learning, and growing that I have done since then have had far more to do with coming to terms with the reality of my existence prior to her death than with coping with her absence since then. I had to get past her life more than get over her death.

A few years after my mom died the TV show Crossing Over with John Edward was at its peak. Before then, I didn't even know what a medium was. I come from a mid-western Catholic family of German descent, so it had just never come up. The show looked legitimate enough. The information reported seemed specific and the people seemed sincerely moved. To this day, I have no reason to suspect otherwise. However, as a scientist, I wanted to see it for myself.

I looked into getting tickets to the show, but the waiting list was years long. While I poked around on the website, I discovered that John (I hope he doesn't mind my unwarranted familiarity) had come to the very university I was attending (UA) and worked with a psychologist there. Through a mutual acquaintance, I was able to get a recommendation from the psychologist for a medium who could do what this guy on TV could do. I got the name and phone number of Angela, a medium in Phoenix, a two-hour drive from Tucson. I held on to the number not really knowing what to do. I went on with my daily life and didn't think much about it. I had, however, told my aunts (my mom's two sisters) about the strange coincidence of the UA psychologist and the medium from TV. A number of weeks later, my aunt Leslie asked if I had called the local medium yet. I hadn't, so I did.

My First (and Only) Mediumship Reading

When I called Angela to make an appointment, I made sure not to give anything away that she could use to mimic communication with my mom later. I had come across a book on the discount table at a bookstore called The Naked Quack: Exposing the Many Ways Phony Psychics &

Mediums Cheat You! by a psychic named Wendy, so I knew a little bit about 'cold-reading' and similar scams. The last chapter of the book detailed the ways in which fake mediums can make it seem as if they are communicating with your deceased loved ones. The book recommended that the sitter not say anything beyond "yes," "no," and "I understand." A good fake medium will change information you say to her into what sounds like novel information and simply report it back to you.

One of the most interesting parts of that chapter addressed a topic entirely new to me (though most mediumship-related issues were). It said that some people think that if you go to a fake medium, believe that communication with the deceased has occurred, and feel better or acquire closure through the experience, then who really gets hurt? Wendy believes your deceased person gets hurt. I agree.

Imagine you are my deceased mother and I go to a fake medium to talk to you. You spend an hour screaming all the things you want to say to me, but the fake medium hears none of it. And not only does she not tell me what you're saying, she makes things up and tells me that you're saying them. At the end of the hour, I mop up my tears, sigh contently, thank the medium for her time, and pay her for the experience. Then I leave, never having heard what you said but believing that I have spent an hour talking with you and that I'll never need to hear from you again. Imagine your pain and frustration. That's who a fake medium hurts. She hurts the dead.

I kept all these threats in mind as I made my appointment with Angela. I silently kicked myself for revealing to Angela that, "Yes, Friday would be fine for an appointment because I already had plans to drive up to Phoenix anyway." 'Oh no,' I thought. 'Now she knows everything! Now she knows I live in Tucson and that I own or at least have access to a car! I've totally contaminated the study!' Needless to say, I was overly cautious about the whole process. Angela and I set up a time and she gave me directions to her home.

I spent the next few days designing an experiment for my appointment with Angela. I figured that I should come up with some sort of code or phrase that my mom could say to Angela so that I would know that it was truly my mother and that Angela and mediumship were legitimate. (No pressure.) But I was very careful (or paranoid) to never actually mention a code or phrase over the phone or over email to anyone because, obviously, this stay-at-home mom could use her many CIA and FBI connections to tap my phone and read my email, having gotten my phone number from her caller ID when I called to make the appointment. (As I said: paranoid.)

I have since written about the problem of using codes and the like in mediumship research. Historically, researchers, philosophers, and skeptics have suggested that, "If a medium could only do [blank], then that would be something!" These 'blanks' have included retrieving the combination to a lock, speaking in a foreign language, and performing a complicated intellectual task using knowledge only the deceased had. However, these so-called 'ideal' protocols make unsupported assumptions about the discarnate's ability and desire to communicate specific types of information as well as the medium's ability to receive and report it. In my paper "Contemporary Methods used in Laboratory Based Mediumship Research," I noted, "Perhaps the discarnate no longer wishes to speak French, play competitive chess, or write a concerto. Maybe without a body constrained by 'earthly' physics, the combination to the lock holds no interest or has been forgotten" (p. 62).

Some of the other proposed examples of ideal mediumship data include interacting with a 'drop-in' communicator unknown to everyone involved and providing information that cannot be fully understood until related information from another reading is obtained (called 'cross-correspondence'). These involve phenomena that usually only occur spontaneously, so (like with near-death experiences) it's hard to design experiments around them.

The other major issue for sitters in the real world (outside of research) is that if you know the answer to the code (or what nickname the deceased called you, the name of your childhood pet, etc.), then it's not that impressive when the medium says it. She could be getting it directly from you. [I discuss this issue in more detail in Chapter 4.] I suggest that sitters don't try to test mediums, that the deceased are quite capable of providing unique and convincing evidence. I didn't know any of this at the time, so using a code during my reading with Angela seemed (in my ignorance) like a good idea.

In the car on the way to the reading, I decided that I should probably ask my mom to participate (this is something we suggest to this day). So just like I did when I was little and wanted my mom's attention when she was clearly busy with something else, I said out loud in the car, "Mom, Mom, Mom, Mom, Mom!" Feeling rather silly, I then said, "I'm going to see this woman who claims she can talk to dead people, so I would really appreciate it if you would go and talk to her while I'm there. And if you do, could you please say something specific so that I know it's you?" I decided on the phrase "Silverton, Colorado." We had vacationed there as a family and my mom really enjoyed it. And it was a phrase that most likely wouldn't come up by chance in a reading.

When I arrived, the medium opened the door to her home and I was surprised at how young and how "normal" she was. Angela was about my age and was not clad in beads or crystals or any other gypsy-like attire. In fact, I looked down at her long denim skirt and said, "I have that same skirt."

Then I silently kicked myself again for giving away such primary information. 'Now she knows where I shop! All is lost!'

I sat on Angela's sofa and she sat in a chair facing me. She explained the different types of readings that she did, psychic vs. mediumship, and I told her I was there for a mediumship reading. She explained how her process worked, what to expect, etc. She then began my first (and only) mediumship reading. This was the only time I have ever made an appointment and received a full reading. Things have spontaneously come up during research, but the deceased members of my family know not to interrupt science, so that hasn't happened in years.

Angela began talking about a male in my generation, she provided the name Ron, that we were teenagers together, that he had died in a car accident, that he was reckless, that he drove too fast, that people had warned him about driving, that he was 17 when he died, that he drove a restored Mustang or other "muscle car," that drinking was involved, that he was aggressive, that he was in my close group of friends, and that he and I joked around a lot. There were a total of 16 pieces of very specific information.

And not a single one was right. Absolutely no friends of mine from high school died in car accidents, male or female. In fact, no one in my whole high school had died while we were in school. And I didn't know anyone who had driven a restored Mustang. At that point, terribly disappointed and frustrated, I thought, 'This is all nonsense. None of this is real.' (I used a much stronger word than 'nonsense' though.) Angela asked if any of that made sense to me. I said it did not and she instructed me to ask around after the reading.

"Maybe this boy came through for someone else you know," she said.

"Whatever, you crazy charlatan," I said—but only in my head. I told her I would ask around about this boy.

Then Angela began describing a woman. She described tucking children into bed and that the overall feeling was that she cared. She described the woman as grouchy and rough but as having humor. She said that this woman held back her feelings, but that she loved her kids. Then she said that it was easier for this woman to place fewer conditions on her emotions for her grandchildren. Neither my sister nor I had children at the time, so I started to think that Angela was describing our maternal grandmother. The information so far definitely described our grandmother and when she added "a genetic tie to Germany" and "she enjoyed accordion music" it became even clearer. I have few memories of my grandmother, but when Angela mentioned accordion music, I vividly remembered seeing an accordion in one of the bedrooms of her home. My aunts later confirmed that my grandmother would demand everyone in the house be quiet while The Lawrence Welk Show was on TV so that she could enjoy the accordion music. (In recounting this story to others I have heard that this was the rule in many families across the US.)

I could neither confirm nor deny a lot of the information about or from my grandmother (including items referencing a small dog, cuckoo clocks, and family China), but a few pieces related directly to me. Through Angela, my grandmother described me looking in a mirror related to her and described herself behind me looking over my shoulder. "Which was funny for her," Angela said, "because she couldn't have done that in life because of her height." I had in my bedroom, and since I was very young, a vanity dressing table with a top that opened to reveal a mirror; it had a matching stool also. It was my grandmother's sister's vanity and at the time of the reading, I sat on that stool in front of that mirror every morning to put on my make-up and do my hair. I am 5'8" so the only way my grandmother could ever stand behind me and look over my shoulder would be if I was seated. (I have since sold the 'haunted' vanity to a local store for grocery money.)

Angela asked if this was the person I was hoping to hear from. When I told her I wanted to hear from my mother, she began again. The following phrases were the first pieces of information Angela reported: "feels removed," "confused," "chemical imbalance," "can't express emotions," "feeling of pain throughout the system," and "whole system attacked." My mother had worked her whole life as a pharmacist and used her knowledge of pharmaceuticals and physiology to end her life. The family of one of her customers had returned the unused portion of a narcotics prescription to the pharmacy for disposal; my mom brought the bottle home and took enough tablets to halt her respiration. The phrases "feels removed" and "can't express emotions" clearly describe the alexithymia from which she suffered; alexithymia is quite literally an inability to cognitively experience and express emotions. Her whole life, my mother was convinced that her problems were physiological, anatomical, or organic in nature. She tried every antidepressant on the market, but never addressed the terribly abusive relationship she'd had with her own mother. Even in her death, the reference to "chemical imbalance" seems to blame biochemistry. The phrase "whole system attacked" appears to describe the effects of the medication on her body's systems at the time of her death.

Later in the reading, Angela used the phrases "ingested pills" and "feeling empty inside, so much that she nauseated herself." She said that my mom "couldn't be prouder" but that she "couldn't express it, she had deep-seated issues." She reported that my mom was saying, "It wasn't you —I was born this way" and that "she was put together wrong." Again, my mother believed for most of her life that her issues could be fixed with drugs or surgery. Toward the end of her life, as her experiences were given a name—alexithymia—and she learned more and more about her disorder, she was just beginning to realize that perhaps when she was very young, her brain and the way it processed information had changed to protect her from a harsh, abusive childhood. It is my personal opinion that she was unable to face this possibility and those memories.

Angela then used the phrase, "She's well." She said that my grandmother had to talk first and show my mom how to access Angela. She then talked about kids pulling themselves up on the base for a stereo from the 1970s with curves, which I remembered (eventually that base was used for potted plants). Angela said that there was a Dalmatian with my mom. "Does that make sense?" Angela asked. "Well, sort of," I said. "Our dog was half Dalmatian and half black Lab." Getting a little exasperated with my hesitancy to assign meaning to anything, Angela said something like, "Come on! Work with me!" Angela claimed that she had never before had a Dalmatian come through in a reading and for several years after that (we stayed in touch) told me that none had showed up since. It was very meaningful and very comforting that this dog showed up with my mom. As with most people, I consider a dog a member of the family, not simply a pet, and that half-Dalmatian never left my mom's side. It is not surprising that they would have found each other on the other side.

Using her background, my mom also delivered some medical advice during the reading. She told me I should "keep an eye on myself in regard to depression" and that she didn't want me to "look at pills as a weakness." That was an issue I struggled with. Even with two parents who were pharmacists and a PhD in Pharmacology, I still viewed medication as a crutch.

Is She Inside My Head?

At one point, Angela talked about a station wagon and our mom wanting to be "a Brady Bunch family" although we were not. I have one vivid memory of that station wagon. It was big and ugly and brown and the third bench seat faced backwards leaving a few inches of carpet between the middle seat and the backseat. I had left a lime green crayon (and not the regular-size kind; it was the big, six-inch kindergarten kind) rolling around on that patch of carpet and the summer sun in Arizona had quickly turned it into crayon soup, at which time it rolled no more. When Angela mentioned the station wagon, I began screaming in my head, "Green crayon! Green crayon!" Angela did not react as if she could hear me and made no mention of the color green or of crayons or any art supplies at all. She just went on with what she was saying. That made me question telepathy (that is, Angela reading my mind) as an explanation for how Angela was getting her information. In addition to the many pieces of information that I didn't know that were later verified by my family, the fact that she didn't report information I was overtly trying to mentally send her eliminated—in my mind at that time—the possibility that she was "simply" reading my mind. [I cover this issue of telepathy in mediumship readings more extensively in Chapter 4.]

More Evidence

During the course of the reading, Angela reported more and more specific, accurate items. She talked about my mom's younger brother. She referenced the month of May (but no other months) which is my mom's birth month (but also Mother's Day). She talked about "headstones in collaboration" tied to my father and that "she walks the grounds there." Ten years before my mom died, my parents chose a cemetery and bought two spots in a mausoleum underneath a beautiful garden. It made sense that she would enjoy the grounds as she chose the location herself. Angela talked about horses and green rolling hills which made her think of Colorado or Montana. I told her about my code phrase being Silverton, Colorado. Angela said that earlier she had seen an image which she, the medium, associated with Silverton. I, like the conservative scientist I am, gave her half credit.

Toward the end of the reading, Angela mentioned the name "Elizabeth." I told her I didn't know anyone by that name. "She's being really adamant about this. It could be a middle name," Angela pressed. "Oh," I grinned sheepishly. "My sister's middle name is Elizabeth." I've found this to be rather common in mediumship readings: Sitters tend to forget the most obvious of facts about their own lives in the heat of the moment.

At that point, Angela had just about had enough of my skepticism, but was very understanding that I was new to all this and didn't really understand how it worked. I thought that a medium is more or less on the phone with the deceased and assumed that all of the information should come through clearly. I had no idea about the intricate symbolism and the interpretation that was involved.

After I spent some time talking with (or more accurately, interrogating) Angela about her mediumship and listening to her stories about her experiences, I thanked her and left. Walking down Angela's driveway back to the car, I felt strangely normal for having just spoken with my dead mother. I decided that the only thing weird about the experience was that somehow it wasn't weird at all. It felt very real and very normal. In the course of those two hours, I was more or less convinced that my mother was still around and that she and my grandmother had spoken to me that day through a nice lady named Angela.

Integrating the Afterlife into My Life

I spent the weekend with some friends of mine from high school who, like me, were very surprised to find out what a normal experience my reading was. I excitedly told them all the things Angela had said. Since I had gone to a different high school our freshman year, I asked them if they knew anyone named Ron who had died or anyone who had died in a car accident. They didn't know of anyone.

When I went back to school/work on Monday, I shared my experiences with the medium with some of my professors and other students. As good scientists, most thought it was interesting and believed that the facts I conveyed reflected what had really happened. A couple, however, were too mired in their cultural, religious, and/or materialist worldviews and literally told me that what I was describing couldn't have happened. As someone with a tremendously strong sense of justice, I thought it was entirely unfair for trained scientists to conclude that a whole phenomenon was impossible even in light of testimony provided by one of their colleagues. I think that it was then and there that my future as a mediumship researcher began.

I systematically scored the reading and found that the percent accuracy was 93% for the items related to my mom. When I asked my aunts to score the reading, they scored items that didn't make any sense to me as accurate. They recognized the small dog that Angela described as being with my grandmother and scored the items "cuckoo clocks" and "China in the family" as related to her. Their scores of the information about my mom were almost identical to the scores I had assigned to those items. The reading was very identifying of my mom.

An Odd Opportunity

I contacted the psychologist who had worked with John Edward to discuss my experience with Angela and the results of the reading. After meeting with him a few times, he offered me a post-doctoral fellowship in his lab. I needed to finish my PhD and then I could start performing research. Oddly, when I was offered the position, I didn't yet have any plans for after graduation, which is terribly unlike me. I constantly have a plan (and usually an accompanying checklist) for most aspects of my life. (I remember my friend Katie razzing me once for having a 'make a list' item on an existing list.) Fortunately, there was no plan and no list preventing me from accepting the offer.

Once it was official, I remembered that I am the biggest scaredy-cat around (I can't watch horror movies or even some TV commercials with strange creatures in them). I figured that a full-time job centered on communication with the dead might just turn my hair white. Much to my relief, it turned out that dead people—at least the dead people I've met—are in no way scary. (If there were lawyers capable of such feats, I would recommend that the dead initiate a class-action lawsuit against the movies and TV shows that have so horribly misrepresented them.)

Advice about My Love Life from the Other Side

Soon after accepting the position, I began dating a man named Corey. When we were younger, our paths had crossed several times, but we'd never actually met. In high school, Corey had been a senior who participated in a lot of school activities when I was a freshman so I had known his name and seen him around. (That school closed that year, so I attended a different school for the remaining three years of high school.) I heard mutual friends refer to him often during the next several years. We even went to the same college (and assume we were at some of the same parties), but never spoke to one another. We finally met at a friend's birthday party a month after my reading with Angela and started dating a few months later.

One evening over tortellini and Chianti, I decided to break it to Corey that I had chosen to take this strange job after I graduated. He had only known that I was a graduate student in the Pharm/Tox department. I began telling him the story about all the amazing things Angela said and about how I had decided to take the post-doctoral position. He thought all of it was really interesting and, like most people, had had enough weird personal experiences to believe that something beyond our five senses is most likely at play.

Then I told him about the beginning of the reading when she told me about the mysterious Ron who had died in a car accident involving a restored Mustang. He had me tell him the rest of the details from the reading and then hesitantly responded with, "Um, I think I know that guy." It turned out that when Corey was a sophomore in high school (and I was still in seventh grade), he had a very close friend named Rick who was a senior. Rick had an aggressive personality, he drove a restored "muscle car," he died driving that car after several people had warned him about his driving, and drinking was thought to be involved; these were all pieces of information the medium had reported. When Corey actually scored the list of items from my reading, he gave the 'Ron' items a total percent accuracy of 93%. When I had officially scored each item as to how they applied to me, the total percent accuracy was 0%.

The implications of Rick showing up in my reading were a little daunting for Corey and me. Did a friend of Corey's show up in my reading, before Corey and I had ever met, knowing that someday Corey and I would be so close that I would share the story with him? Granted, I did go to the same high school as Rick, but he died two years before my first day of school and I never even heard the story or his name. It's a heavy load to deal with at the very beginning of a relationship. Did it mean that Corey and I were destined to meet and start a relationship? Did it mean that Rick had been hanging around Corey and/or me and/or our high school since he died? Or just since we were far enough on our paths that he knew we'd meet? Did he just want to say a hello to Corey or did his presence in my reading provide evidence for survival of consciousness and the interconnectedness of all things in the universe?! Corey and I chose not to dwell on the universal implications and just tried to have a relationship based on normal (vs. paranormal) commonalities. For two years, that was the case.

University Research

After getting my PhD, I started performing mediumship research in June of 2003. I was the first (and, sadly, last) William James Post-doctoral Fellow in Mediumship and Survival Research at the UA. I also served as co-director of the VERITAS Research Program there. My position lasted for four and a half years and was funded solely by one generous individual for that entire time.

As the primary hands-on researcher at VERITAS, I managed the day-to-day activities of the research program (for some of the time, I even lived in a different city). My responsibilities included hypothesis development; protocol design; consenting, screening, and training human participants (the mediums); conducting experiments; data collection, entry, and analysis; manuscript writing and editing; and managing undergraduate student research assistants. I was the researcher on the phone with the mediums during the study readings and I pored over all the data, so I really got to understand the intricacies of the phenomenon. The culmination of that research was a study that addressed the primary question of mediumship research: Can mediums report accurate and specific information about the deceased?

When the funding for my position was nearing its end, I tried all kinds of ways to keep the research going. Unfortunately (or fortunately, I guess), metaphorical door after door (for example, fundraising, the possibility of a different position within the University, etc.) was kindly closed in my face. It seemed that the universe had left only one path open: go out on my own. Once I started to pursue that course of action, support (mostly emotional) seemed to spring up everywhere. I co-founded an organization with my husband Mark Boccuzzi in order to continue performing mediumship research: The Windbridge Institute for Applied Research in Human Potential. Soon, we had a website, a logo, and an extensive group of renowned scientific advisors. The following year, we even received a grant to support a research study.

The rest of this book describes what we did and what we discovered.

But first...

In an attempt to establish myself as an honest-to-goodness real person and not some kind of single-minded science machine, I will now share with you some random bonus facts about myself (that might come up if we were, in fact, having this conversation over coffee):

I have two tattoos (so far).

To remain happy and healthy (mostly healthy), I sleep 10-12 hours per night (and I highly recommend it).

The center of my universe is a 13-year-old, 90-pound, second-generation rescue-American mutt (her parents were both rescued, but she was never in a shelter) named Moose (my husband recognizes the pecking order—Moose came first).

I am a huge Denver Broncos fan (thanks in part to two different boyfriends).

I know lots of interesting facts about the human body (like how you should push your tongue against the roof of your mouth to treat 'brain freeze'—it's cold sinuses that cause the pain and your tongue helps warm them up).

I subscribe to an online comics service and end most days by smiling at those just before bedtime.

Although I am a practicing vegetarian, I hate vegetables (I'm more of a fruitarian).

I am not ticklish.

Julie Beischel, PhD

❦ 3 ❧

Science for All

*Science is a way of thinking
much more than it is a body of knowledge.*
—Carl Sagan

At my very core, I love science. It is the tool with which I interact with the world each moment of each day. What is that noise coming from the kitchen? I think (that is, I hypothesize) that it was the dog food bowl, but I better check it out. It sounded metallic in nature so that rules out the plastic microwave cover falling into the sink. My surveillance of Moose (the dog) upon entering the kitchen leads me to conclude that she wasn't concerned by the noise and allows me to eliminate as an explanation that it was a foreign or threatening noise like a burglar. After observing that all metallic items are in their normal places (that spoon is on the drying rack where I left it and my keys are still in their dish), I note that Moose's metallic food bowl is, indeed, on the floor. I conclude from my investigation that, in licking every last molecule of dinner out of the bowl, Moose has licked it right on to the floor. Hypothesis confirmed.

Welcome to my world.

I do, however, understand that not everyone moves through the world that way. Also at my core, I am an educator. I know that if I can't get a student to understand what I am trying to convey, that is my problem, not hers. I have experience teaching complex scientific concepts to regular people. I even once taught RNA transcription and translation using the analogy of a sandwich (the glycosylation was the olive on the toothpick). So please rest assured; I will get you through the following discussion of the scientific method unscathed.

Mediumship and the Scientific Method

As with any natural phenomenon, the traditional scientific method can be readily applied to mediumship. Let's walk through that process:

The first step in the scientific method is to make an observation. In this case, our observation is: That lady claims she can communicate with the dead and words related to deceased people are coming out of her mouth. Check.

Step 2 is formulating a hypothesis about the observed phenomenon. A hypothesis is usually a positive statement (that is, "this happens") rather than a negative (that is, "this doesn't or can't happen"). The negative statement is called the null hypothesis and is the opposite of the stated hypothesis. In my previous life in physiological research, a hypothesis would have been "This drug decreases blood pressure" or "Cell type X in the immune system is what causes the damage during Disease Y." In the case of mediumship, the hypothesis might be: That medium is talking to the dead. But because it hasn't been proven yet that the dead are any more than bodies in the ground, that statement gets a little ahead of ourselves. We will start with: The statements the medium is making about the dead are specific and accurate.

The third step of the scientific method involves designing an experiment that tests whether the hypothesis is true or not. The ideal experiment examining mediumship includes two important factors. Without these factors in place, we really won't know anything more about mediumship after the experiment than we did before it. The key is finding the right balance—the sweet spot—between them.

The first of those two important factors is an environment that mimics how the phenomenon exists 'in nature.' As with the study of any phenomenon, mediumship research should focus on where and how the phenomenon exists in reality. Creating an experimental environment too far removed from the natural mediumship process would be like placing an acorn in your palm, waiting a few minutes, and then calling it a fraud when it didn't turn into an oak tree. Like the acorn, mediums need the equivalent of soil, water, and sunlight to effectively do what they do.

When designing our mediumship experiment, we need to use a setting similar to how a medium performs a reading in her everyday practice. We can't ask her to report information about Napoleon while standing on her head. That's not what she does. In her practice, she reports information about deceased people (whom we call discarnates) to their living loved ones (whom we call sitters). So, first we need to include the sitters in the experiment. We also need to ensure that we only ask the medium to report the types of information she usually reports. Since this does not include winning lottery numbers, combinations to locks, or what color shirt the sitter should wear tomorrow, we won't ask for any of those things.

We also need to provide the medium with a nugget of information that she can use to focus or 'connect.' In a regular, non-research reading, the medium may use the name of the discarnate, his or her relationship to the sitter, or the mere presence of the sitter during the reading in order to focus. In our experiment, we'll give the medium the first name of the discarnate.

Another piece of the optimal experimental mediumship environment involves testing good mediums. If we wanted to study the phenomenon of high jumping, we would find some good high jumpers. We wouldn't invite some people off the street into the lab and tell them, "Go jump over that bar." When those people couldn't do it, we wouldn't have learned anything at all about high jumping. In mediumship research, we should select participants with a track (and field) record of reporting accurate information about the deceased.

We should also be careful about who does the accuracy scoring. The information reported in a mediumship reading is a personal conversation between two people with an emotional connection: the discarnate and the sitter. Even if I asked you to tell me everything there is to know about your deceased loved one, truly meaningful information may still come up in the reading that you hadn't thought about in years. If I tried to score that information based on what I had collected from you, I would claim it was inaccurate, but that claim would be wrong. Only you can decide what is identifying and accurate about your discarnate. This issue may have been relevant as you read Chapter 2. You may have thought that the information Angela provided during my reading was general or unimpressive. However, to be blunt, it doesn't matter what you think; it was a reading for me, not you. Only the people who were close to my mother and my grandmother are qualified to assess the accuracy and meaning of that reading. Thus, in our protocol design, we should only be concerned with accuracy scores provided by sitters. That's how it works in a regular reading.

The second important factor we need to make sure to include in our design of the experiment is the use of proper experimental controls. We need to eliminate all the normal explanations for how the information the medium reports could be accurate. To rule out fraud, we have to make sure the medium can't look up information about the sitter or the deceased person online or in any other way. We also need to account for cold reading, a process where cues from the sitter or leading questions are used to steer a reading so it seems as if it's accurate. To prevent that from happening, the medium will be what's called masked or blinded to the sitter. The medium won't be able to see, hear, smell, etc., the sitter during the reading; but, as stated above, the sitter should be involved somehow in order to optimize the environment, so we'll just make sure his intention is that his discarnate communicates with the medium. Now if I as the experimenter know things about the sitter or the discarnate during a reading, I could also cue the medium. For example, if I knew the discarnate's cause of death and the medium reports a sharp pain in the head, I could mislead her by saying, "Oh, so the cause of death is breast cancer. Got it." So in our design, let's also blind me to information about the sitter and the discarnate.

With all the blinding, the medium can't get specific information through normal means, but she could just say things that are so general they could apply to anyone like "There's an 'M' name somehow connected to this person," or "She's making a reference to a special meal on a holiday." This is also a form of cold reading. To deal with that, we'll ask the medium specific questions so there's no way the information can be general. We will ask the medium to describe the discarnate's physical appearance, personality, hobbies, cause of death, and messages for the sitter. The use of these questions will further address any general information she could guess from the discarnate's name. For example, it would not be possible to accurately guess the hair color, eye color, height, weight, personality, activities, and cause of death of discarnates named Jennifer, Barbara, Anna, Linda, James, Michael, Jessica, Margaret, Brian, Joe, Nicholas, or John (actual study discarnate names) or the personal messages those discarnates would have for their sitters.

In our design, that just leaves the sitter. When a person reflects on the accuracy of a mediumship reading that he knows was intended for him, his personality and psychology affect how he rates the statements. A person who is more laid-back and forgiving may score more of the items as accurate whereas someone more cynical and strict may only score a few as right. That phenomenon is called rater bias. If we give a sitter more than one reading to score, and only one was actually provided by the medium for him, his bias will be spread evenly across the readings. That is, a cynical person will score all readings with a severe eye and a more tolerant rater will give out more high scores across the board. To maintain blinding, the sitter won't be able to tell which reading is which. We could just look at what people score their own "target" reading as compared to what they score other "decoy" readings not intended for them. That would also further address the generality issue. If a medium reports very general information, all the sitters will score all the readings as accurate.

So, to account for fraud, cold reading, experimenter cueing, general statements, and rater bias, we will need to design an experiment in which the setting is similar to a normal mediumship reading but where the medium, the sitter, and the experimenter are all blinded. The remaining steps of the scientific method are to perform the experiment, draw conclusions from the data, and then start over based on the new observations.

If you understood all of those basics about mediumship research and the importance of optimal environments, maximum controls, and skilled participants, congratulations! You are now qualified to assess the claims of people asserting that they are performing useful scientific mediumship experiments or demonstrations. Let's have a pop quiz!

If the experimenters ask the medium to perform five readings in five and a half hours alone in a room, has an optimal environment been established?

       Answer: No!

If the experimenters choose the mediums for their study from the phone book, have the skills of the participants been scientifically established?

Answer: No!

If an experimenter puts a screen in the room between the medium and the sitter, shows the medium a picture of the discarnate, or has specific information about the discarnate or the sitter, has maximum blinding been ensured?

Answer: Not even a little!

Great job! All of those are real examples; some of them were even published in peer-reviewed journals. So be wary of claims of "scientific" endeavors even if they include fancy words like 'hypothesis,' 'analysis,' 'research,' and 'statistics' and even if the claims are being made by people employed by a university. Now you can make your own assessments about the validity of such claims.

When I applied the scientific method to the phenomenon of mediumship using optimal environments, maximum controls, and skilled participants, I was able to definitively conclude that certain mediums are able to report accurate and specific information about discarnates without using any normal means to acquire that information.

The End.

Just kidding. Let's delve a little deeper into the actual research.

Mediumship Research at the Windbridge Institute

As stated above, one of the most important issues in performing mediumship research is working with skilled mediums. At the Windbridge Institute, we have used an extensive screening, training, and certification procedure to acquire a team of nearly 20 Windbridge Certified Research Mediums (WCRMs). The certification procedure consists of eight steps in which the prospective WCRM is interviewed, tested, and trained. During the testing, the medium performs readings under various blinded conditions and if her accuracy scores achieve a certain level, she goes on to complete the training steps which involve her learning about the regulations governing research with human subjects, the history of modern mediumship research, and grief. Upon completion of all the steps—which takes several months, at least—the medium is certified a WCRM. A full description of the screening process is available on pages 50-56 of my article "Contemporary Methods Used in Laboratory-Based Mediumship Research" published in the Journal of Parapsychology.

Each WCRM donates at least four hours per month to research and agrees to a specific code of ethics which includes confidentiality regarding the content of readings and not performing readings outside of those specifically requested (i.e., not offering unsolicited readings) as well as general good citizenship practices. This group of WCRMs is part of our research team. They assist in protocol development, participate in research readings, and perform demonstrations during public events. They are willing—for the good of science—to attempt experimental protocols that go well beyond their comfort zones and, in an upcoming hematological and psychophysiological study, some of them are even willing to let me poke them with needles to draw their blood. Therefore, we want them to be kind, honest, trustworthy, compassionate, humble, and respectful: the kind of people we want to be around (a characteristic that many mediums I have met do not share). Mediums who do not follow the ethics guidelines are (and have been) removed from the program. The WCRMs can also serve on specific committees if they have particular interests or skills; for example, physiology, ethics, and/or forensics.

I will now try to answer the most frequently asked questions about the WCRM certification procedure:

About 25% of the mediums who have attempted the testing did not achieve passing scores. This does not mean they are not good mediums; it only means that the readings they performed on those days, with those discarnates and sitters, under those conditions did not achieve the level of accuracy we require. On a different day with different discarnates and sitters, who knows?

Roughly 90% of the mediums (prospective and certified) are female. I think this may accurately reflect the general American population of practicing mediums. This is why, when referring to a medium in general in this book, I use "she" and "her." I also use the male pronouns to refer to sitters. It just makes it so much easier to read than "s/he" and "him/her." I officially apologize to all male mediums (especially our two male WCRMs) and all female sitters for this text shortcut.

No money changes hands as part of the WCRM-Windbridge Institute relationship; that is, we do not charge a medium to be tested and we do not pay them for the readings and other research-related tasks they perform.

Our WCRMs are located all over the US. All of our study and certification readings take place over the phone; I've never even been in the same room as about half of the WCRMs on our team. Phone readings are just as good if not better than in-person readings, so a medium does not need to be in your area in order for you to receive a good reading.

We are no longer testing new mediums. The reasons for this are numerous.

(a) The screening is an extremely time- and resource-intensive procedure. It can take up to a year to complete and costs us between $7,000 and

$10,000 per medium (depending on how many sitters need to be screened to find two that qualify). The current WCRMs were screened as part of projects funded by grants; at this time, we simply do not have the time or personnel to test anyone new or to retest the mediums who did not pass on their first attempts.

(b) The number of WCRMs we work with is plenty in order to answer our current research questions. The WCRMs serve as a sample of the larger population of US secular mediums. Like with almost any other research topic, it is impossible to study every single example of the phenomenon under investigation, so scientists study a sample of the population and extrapolate their findings as descriptive of the whole group. I can't work with all of the mediums in the world, so I am currently working with a sample of mediums. That sample consists simply of the mediums who contacted me and passed a comprehensive screening procedure first.

(c) Lastly (but perhaps most importantly), we are in the business of performing cutting-edge research, not in the business of certifying mediums. Even if we charged a fee to the mediums being tested, that's all we'd have time to do and we couldn't perform any actual research (what good are research mediums with no research in which to participate?). We are, however, currently developing studies that will involve a larger population of unscreened mediums.

There are three main arms to the mediumship research program at the Windbridge Institute. Their code names are Information, Operation, and Application. The Information Studies examine (you guessed it) the information the mediums report; specifically, its content and accuracy. This has also been called proof-focused mediumship research. The Operation Studies focus on the process of mediumship (process-focused research) including the mediums' (a) physiology, (b) psychology, and (c) experiences; this research of a person's experiences as well as the person's experience of experiencing the experience (I know, right?) are both called 'phenomenology.' The third arm, the Application Studies, is concerned with the practical social applications of mediumship readings. We'll talk about each of those research programs in turn in the following chapters.

4

Information Research Program

*There are things known
and there are things unknown,
and in between
are the doors of perception.*
—Aldous Huxley

The standard reading protocol we use in our mediumship experiments at the Windbridge Institute accounts for all of the normal explanations for a medium's accuracy discussed in Chapter 3: fraud, cold reading, experimenter cueing, general statements, and rater bias.

A typical experiment involves one WCRM, two sitters and their respective discarnates, and three experimenters. It goes a little something like this...

Sitters and Discarnates

Over 1,000 people have volunteered to serve as sitters in our mediumship research by completing an online form. At the start of an experiment, one of them is chosen at random and then software we developed (and by 'we,' I mean my husband Mark and our research volunteer and software guru Ryan) is used to pair that randomly-selected sitter and their primary discarnate—the deceased person the sitter would most like to hear from—with a second sitter and their discarnate. The pairing specifically matches the first discarnate to his or her opposite. That is, the deceased people are matched to be different in their physical and personality characteristics, their interests or hobbies, and their causes of death.

We perform this pairing because at the end of the experiment, a sitter scores two readings for accuracy—one that was intended for him (a target) and one that was intended for a different sitter (a decoy)—in order to account for rater bias. Now if both of those readings described, say, young men in their 20s with brown hair, green eyes, and slender builds, who liked video games and playing guitar, and died in traffic collisions involving motorcycles, it would be nearly impossible for a sitter to distinguish which reading was for his son. And that could happen if two discarnates were paired randomly; people are only so different. In that situation, we really wouldn't have learned anything about mediumship at the end of the study.

To begin an experiment (I should warn you that this part gets a little complex), Experimenter 1 (E1) uses the software to pair two discarnates who are different and then consents, screens, and trains the associated sitters. Mark usually serves as E1 with the assistance of a research volunteer and told me that that whole process involves nearly 50 separate emails per discarnate pair. In the meantime, Experimenter 2 (E2; that's me) schedules two readings with a randomly-selected WCRM. After the sitters are fully screened, E1 emails the first names of the two discarnates to E2 who then randomizes their order and notifies E1 of the scheduled times that each of the readings will take place. (And that is how we refer to each other at the dinner table. "Please pass the salt, E1." "Here you go, E2.").

E1 remains blinded to which WCRM will perform the readings but notifies each sitter of his reading time so that he can request that his discarnate finds the right WCRM and communicates with her at that time. We don't know how that works, but it seems that the medium does not contact the correct, say, Jack; it is Jack who appears to find the correct medium. Jack may even begin communicating with the medium before I even call her to start the reading. In the protocol, we simply instruct the sitters to request that their discarnates participate; we suggest doing this silently like a wish or a prayer. [An interesting story about how it appears that the discarnates are in control of the experiments is available in my blog post "The Two Sally's."]

Research Readings

At the time previously scheduled with the WCRM, I (E2) call her on a cell phone which I have attached to a digital audio recorder. After confirming her permission to record the call, I give her the first name of one of the discarnates in the pair. That is the only information about the sitter or the discarnate that the WCRM or I have. Once the WCRM has taken a few moments to connect with the discarnate, I ask her a standard set of questions:

1. What did the discarnate look like in his/her physical life?

2. Describe the personality of the discarnate.

3. What were the discarnate's hobbies or activities? How did s/he spend his/her time?

4. What was the discarnate's cause of death?

5. Does the discarnate have any comments, questions, requests, or messages for the sitter?

6. Is there anything else you can tell me about this person?

At the time of the second scheduled reading (often the following week), we repeat this process with the second discarnate's name. I then transcribe the recordings to create two lists of numbered items. I remove all references to the discarnates' names and all qualifiers like "I'm seeing..." and "I think that..." so that all that's left are definitive statements like "brown hair," "she was intellectual," "the image of a swimming pool," and "the feeling of an impact to the chest."

Next, I email the two lists to Experimenter 3 (E3; usually our research assistant Mike) and E1 emails him the sitters' contact information. E3 then emails the two readings to the two different sitters and each sitter scores the reading for his discarnate (the target) as well as the reading for the other sitter's discarnate (the decoy) without knowing which is which. This scoring usually involves each sitter:

(a) providing accuracy scores for each item in each reading,

(b) giving each reading an overall global score from 0 to 6, and

(c) choosing which reading he thinks was intended for him.

This fully-blinded protocol includes five levels of blinding:

1. the WCRM is blinded to information about the sitter and the discarnate before, during, and after the phone reading;

2. the sitters are blinded to the origin of the readings during scoring; they do not hear the readings as they occur;

3. the experimenter who consents and trains the sitters (E1) is blinded to which mediums read which sitters and which item lists were intended for which sitters;

4. the experimenter who interacts with the mediums during the phone readings and formats the readings into item lists (E2) is blinded to information about the sitters and the discarnates beyond the discarnates' first names; and

5. the experimenter who interacts with the sitters during scoring (that is, e-mails and receives by e-mail the blinded readings; E3) is blinded to all information about the discarnates, to which medium performed which readings, and to which item lists were intended for which sitters.

More detailed information about the methods used in the Information arm of research at the Windbridge Institute is available in my article "Contemporary Methods Used in Laboratory-based Mediumship Research."

Results

At this time, we have scoring data from 21 readings. The sitters scored more of the items in their own readings as accurate versus the number of items they considered accurate in readings intended for other sitters and this difference was statistically significant. For example, you might score a reading for someone else's discarnate as 30% accurate and a reading for your own discarnate as 65% accurate. Even in an example, those numbers are far from 0% and 100%, respectively. First, like I said above, people are only so different and so some items will apply to each discarnate in the pair.

Second, mediums, especially under quintuple-blinded conditions where there is no real-time connection between the sitter and the discarnate, are far from 100% accurate (regardless of what you've seen mediums on TV do). That's only logical; it's not a phone call with the dead. Even in an unblinded reading, there are multiple levels of interpretation potentially involved: the discarnate may be unable to convey specific types of information, the medium may misinterpret some of the symbols, and the sitter may misinterpret certain items reported by the medium. And like with any job, mediums have their good days and their bad days. Be wary of mediums who report their accuracy percentages. Upon encountering these claims, you may want to ask about the blinding conditions of the readings during which the accuracy rating was achieved, if the reported number is an average or a one-time best, and what information the medium had about the discarnate(s) and the sitter(s) prior to the reading(s).

We also found that sitters gave their own target readings higher overall scores than they gave to the decoy readings (again, to a statistically significant level). Finally, 16 out of 21 (76%) of sitters chose their own target reading as the one they felt was intended for them when asked to choose between the two readings without knowing which was which. If the sitters were choosing purely by chance, we would expect 50% to pick the right reading. At 76%, this data is also statistically significant. In a scenario like this, the statistical tests 'understand' that when you flip a coin 21 times with a 50-50 chance of getting a head and you get 10, 11, or even 14 heads, it's not impressive. Now, if you get 16 heads out of 21, that's something!

What does all this Mean?

These data collected as part of the Windbridge Institute Information research program support the initial hypothesis: Certain mediums can report accurate and specific information about discarnates with no prior knowledge about the sitters or discarnates, without any feedback during the reading, and without using fraud or deception.

We refer to this phenomenon as Anomalous Information Reception (AIR). There's no normal way the mediums could acquire the information they report so its reception can only be described as anomalous.

Here's the rub: we cannot conclude from this data that the mediums are communicating directly with the deceased. Though they are not acquiring the information they report through normal sensory means, there are still three competing suprasensory explanations for how the information might be acquired. I'll now go through each of these.

Life after death

In the first explanation, the mediums are actually communicating with the deceased who continue to exist, sans physical body, in the afterlife. This theory is officially called the survival of consciousness hypothesis or just 'survival' for short and states that a person's identity, personality, consciousness, or self continues to exist after the physical death of the body. In the survival explanation, mediums are, in fact, communicating with the dead. But like I said, this is just one possible explanation.

Psychic reservoir

The psychic reservoir hypothesis claims that all the information since the beginning of time is stored somehow and somewhere in the universe (like the Akashic field/record) and mediums are accessing that cosmic store of information to acquire facts about the deceased.

Super-psi

Psi (not PSI or P.S.I.; it's the Greek letter psi, ø) is an umbrella term which includes several different types of anomalous cognition phenomena. Historically, these individual cognitive phenomena have been called telepathy (the acquisition of information from someone else's mind), clairvoyance (the acquisition of information from a distant place), and precognition (the acquisition of information from the future) and collectively termed extrasensory perception (ESP). However, because of very loud tirades from people who flatly deny that these phenomena exist (even though there is more scientific evidence for the existence of psi than there is for the role of aspirin in preventing heart attacks), these terms are emotionally charged and make many people itchy for some reason. So we will stick with neutral, bland terms like 'anomalous cognition' and 'non-local information exchange' ('psi' for short). [Note: Officially, psi also includes mind-matter interactions or psychokinesis, but here we'll use the term psi to refer to the mental ESP phenomena only.]

In the super-psi (or super-ESP) explanation for where a medium gets her information about the deceased, she uses psi to acquire information from the minds of the sitter, the experimenter, or others; from a future time when she receives feedback about the reading; and/or from distant items, locations, photos, documents, etc. Because super-psi is basically all-encompassing (thus the 'super'), it cannot be disproven (any more than could a theory proposing that God is whispering the information to the medium or one in which aliens are beaming the information directly into her head) and, therefore, does not actually qualify as a scientific hypothesis. It is more of a philosophical tool for discussing alternatives to survival.

Who's right?

So which of these three explanations best fit the Information data? They are all neck-and-neck. There is no evident leader. It is impossible to distinguish between survival, psychic reservoir, and super-psi based on the content of mediumship readings (or any information allegedly from the Great Beyond). We could collect accuracy data until the end of time and it would continue to support the existence of anomalous information reception by mediums, but it could not help us get any closer to figuring out where a medium gets her information. The brick wall at the end of this 'survival vs. psi debate' has been plaguing mediumship research for more than a century and is one of the main reasons why this line of inquiry is not embraced by more researchers (you know, in addition to the academic mockery, limited grant opportunities, and lack of employment options).

Now, I know what you're thinking. "What about when a medium reports information the sitter doesn't know and needs to be verified by other people or looked up? What about when she reports something that hasn't happened yet and then it does happen? Those are clearly evidential of survival." Nope. And nuh-uh. Those phenomena can also be explained by psi (telepathy/clairvoyance and precognition, respectively). The content of the reading is irrelevant in this debate because it can never break the tie or shift the balance. So we will continue to be stuck at this impasse, this deadlock, this stalemate.

The End.

Just kidding again.

At the Windbridge Institute, we did a crazy thing to get to the bottom of this issue that no one had done before: we asked the mediums what they thought.

To be fair, historically, mediums entered into a trance state during their readings and had no conscious experience of the communication so they couldn't comment about it and most investigators concentrated on the content and accuracy of the readings in order to establish the existence of AIR, so it wasn't really an option to bring the mediums into the debate.

However, with the scientifically-minded WCRMs, we were able to get their input on the psi stalemate. You know what they said? "We know what psi feels like and this ain't it." (Though no WCRMs would actually use the word "ain't.")

So we embarked on the second arm of mediumship research at the Windbridge Institute—Operation— and systematically examined their experiences. The next chapter discusses what we did and what we found.

Julie Beischel, PhD

❧ 5 ❧

Operation Research Program

*Nothing in life is to be feared.
It is only to be understood.*
—Marie Curie

The Operation mediumship research program at the Windbridge Institute involves studies of mediums' experiences (phenomenology) as well as potentially unique characteristics of their physiology and psychology.

Phenomenology

Skeptics will claim that it behooves mediums to experience information about the living and about the deceased as different since no one would pay them to get information about rather than from a dead loved one. (I have a colleague who dismisses our phenomenology research outright because he once encountered one 'medium' who said she only pretends to receive the information she reports as coming from the dead. That wasn't a medium; that was a psychic fraudulently posing as a medium.)

Julie Beischel, PhD

Those skeptical individuals (who often have never even heard a mediumship reading or have only encountered fraudulent mediums) will claim that mediums may be (at best) subconsciously deluding themselves into experiencing the two types of information differently or (at worst) actively lying about it.

The truth is, our very basic sensory perceptions are not any more directly reflective of reality than a photograph is directly reflective of the physical environment (seeing is the brain's best guess), so being skeptical about anyone's claims about the validity of their perceptions is logical and warranted. For example, when a person hears words or sees a face, it only means a human brain was involved in the perception (the first thing an infant brain learns to hear is a voice and the first thing it learns to see is a face); it doesn't mean that objective voices or faces exist. And even if 1,000 people see the face, it only means 1,000 human brains did the seeing. Perception is flawed. (For more on the fascinating world of how what we 'see' is based on the least amount of brain resources required to keep us from being eaten by predators, see Visual Intelligence: How We Create What We See by Donald D. Hoffman.)

Thus, I don't fault people for being skeptical of mediums' claims regarding their experiences. However, those claims start to become rather trustworthy when taken together with (a) at least some mediums' experiences having begun when they were only small children with no financial or other ulterior motives and (b) at least some mediums' abilities to distinguish between living and deceased targets under blinded conditions (discussed below). Further systematic research was required.

In our phenomenological research, we turned to the mediums' experiences in order to address the historical survival vs. psi impasse. There are several spontaneous phenomena that occur during mediumship readings that tend to be more evidential of survival than of psi. These include mediums being 'corrected' by discarnates during readings and being surprised by the information they receive (which sometimes results in the giggles). We wouldn't expect these types of phenomena if the medium was using psi to reach out to acquire the information, but they are indicative of communication (I like to say that we study human communication; one of the people just happens to be dead). This difference between "retrieve" and "receive" (or dead information vs. dead person) helps us begin to lean in the direction of survival as the best mediumship explanation, but these phenomena are spontaneous and cannot be designed in to research protocols ("Now be corrected by the discarnate" or "Now find what the discarnate is communicating funny").

There are systematic ways, however, to scientifically investigate people's experiences. I worked on several of these phenomenological studies of mediums' experiences with Adam Rock, my colleague at the Phoenix Institute of Australia, and we were able to demonstrate a number of unique properties of the mediumship experience.

In one study, the WCRMs (there were seven at the time) completed a questionnaire about their experiences after a mediumship reading for a discarnate as well as after a regular phone conversation with me (the control condition). The results showed that the mediums experienced higher levels of negative emotions as well as alterations in sense of time, body image, perception, and general state of awareness during the readings compared to the regular phone call. Conversely, they experienced lower levels of self-awareness, volitional control, and memory. These results were in line with what I had observed anecdotally. A medium may take on the emotions of the relationships between discarnates and sitters during readings and this often results in the medium feeling grief, loss, anger, remorse, regret, and/or other negative emotions. The reading condition also seems to change the mediums' experiences of the passage of time (they are often surprised by the amount of time that has passed since the start of a reading) and their physical ailments (headaches, allergies, etc.) often abate during readings. They also often have difficulty remembering the content of some of the items that they just reported. The finding of lower volitional control in the mediumship reading condition is the one that speaks most to the survival question.

This finding demonstrates how the information comes in to the medium rather than the medium reaching out to acquire it. [The journal article describing this study is titled "Quantitative analysis of mediums' conscious experiences during a discarnate reading versus a control task: A pilot study."] More studies were required to fully address this issue.

Most of the WCRMs also perform psychic readings for the living as part of the services they offer to the public, so they have experience with acquiring information about the living using psi as compared to their experiences of communication with the deceased. In one study, we asked the WCRMs (there were six by this time) to describe in as much detail as possible their experiences (a) communicating with discarnates and (b) performing psychic readings for the living. The order of the questions was randomized for each WCRM and a week passed between their answering of the two questions in order to address the impact that answering the first question might have on the second. A third researcher who was not familiar with mediumship evaluated the sets of responses and found some similarities (as expected—it's all extrasensory information) as well as some differences. In both mediumship and psychic readings, the WCRMs experienced the information simultaneously through multiple "sensory" modalities. A medium might see images in her mind's eye as well as hear, feel, smell, and taste stimuli in her mind. For both types of readings, the mediums also reported the experience of "just knowing" certain items.

In just the mediumship readings, the WCRMs reported the presence of signs that contact had been made. These signs included sounds like rings or whines, light flashes, or feelings of vibrations or heat. They also reported the mediumship experience as involving independent, autonomous communicators who could surprise and even frighten the medium with their presence and who had opinions with which the mediums didn't always agree. Another difference between the mediumship and psychic reading descriptions was that while the WCRMs actually experienced the emotions of the discarnates during mediumship readings, they were merely aware of the emotions of the living clients during psychic readings. (The journal article describing this study is titled "Psi vs. survival: A qualitative investigation of mediums' phenomenology comparing psychic readings and ostensible communication with the deceased" and is full of quotes from the WCRMs' responses.)

Through this study, we demonstrated how mediums experience communication during a mediumship reading as quite different from a psychic reading in which psi is used. Perhaps the quote that most succinctly describes this difference is this: "A psychic reading is like reading a book... A mediumship reading is like seeing a play."

But we're not finished yet. Because the WCRMs know which type of reading they are performing in each case, skeptics will bring up the alleged psychological motivation (even if it is subconscious) to experience the two as different, again, because no one would pay them to talk to their deceased loved ones if they were only using psi to acquire static information from a psychic database. Also again, the fact that many mediums have their initial experiences of both psi and discarnate communication at an age so young they have not yet classified the experiences (much less started seeing clients) tends to refute this claim. Regardless, the next step is to examine mediums' phenomenology during the two types of readings under blinded conditions.

In our current study, we have the WCRMs perform readings under our standard fully-blinded protocol and some of the first names they receive are living people and some of the names they receive are deceased people. They (and I) don't know which are which and we perform readings in which I ask them basic questions about the physical and personality characteristics and interests of the named target. After each reading, they fill out a questionnaire about their experiences. This study is currently in-progress, but I will tell you that, so far, in 83% of readings, the WCRM accurately reported the status of the target (living or deceased) which is statistically significant. That is, when only given the first name of a target, WCRMs could tell if the person was living or dead

83% of the time. [To be notified when a journal article describing this study is published, please join our email list by visiting www.windbridge.org]

This is still not definitive evidence solving the survival vs. psi debate, but it sure helps us lean in the survival direction. At this time, we can draw the following conclusions from the phenomenology leg of our Operation research arm (an arm with legs?—gross).

WCRMs make a distinction between discarnate communication and the use of psi during psychic readings for the living.

They experience discarnate-associated information as coming from independent, volitional beings separate from themselves and not as knowledge obtained from a dormant, non-living source like the psychic reservoir or through the psi processes they use during psychic readings for the living.

It is possible that a combination of all three proposed explanations (that is, survival, super psi, and psychic reservoir) may be involved in where a medium acquires her information. The existence of psi abilities and/or the presence of a cosmic store of information do not preclude the possibility of survival. (There can be an Akashic field and ghosts!)

Whatever is going on (though I'm leaning toward survival), it is interesting and worthy of further investigation. In our Operation Research Program, we are also interested in examining the physiology and psychology of mediums.

Physiology

Studying mediums' physiology (how their bodies work) is relevant and important for two main reasons. First, the results may help in predicting, preventing, managing, and/or treating medical issues in mediums. When I informally queried the current WCRMs about any health issues they had, I noted that chronic medical problems may be a serious concern for this population. For example, they have seven times the incidence of autoimmune disorders compared to the incidence in the general US population. Their incidence of diabetes is nearly twice the national prevalence. And the incidence of migraines in female WCRMs is nearly two and a half times the prevalence in women in the US. This is just a small sample, but examining mediums' physiology may uncover the underlying biological causes of these health issues and, in turn, lead to their management in the larger population.

Second, a more comprehensive understanding of the mediumship process will bring it further into the realm of respectable scientific inquiry and experience. Demonstrating true physiological changes may help show that mediumistic communication is a normal human process with normal human physiological correlates. Likewise, documenting a biological basis for the mediumship process will begin to remove the social stigma associated with mediumistic experiences in some communities. This is especially important in order to limit detrimental and unwarranted pathological clinical diagnoses in young children first experiencing these phenomena. In addition, demonstrating that mediumship abilities are common, normal, and healthy may help dissuade parents trying to get their 'special' children on TV and, instead, encourage them to concentrate on helping the children cope with these particular experiences.

In 2011 and 2012, I worked on a physiology project with Arnaud Delorme and Dean Radin, researchers at the Institute of Noetic Sciences (IONS), and Paul Mills from the University of California, San Diego (UCSD). With funding from the Bial Foundation in Portugal, we were able to collect electroencephalograph (EEG) and other physiology data from several WCRMs. I received another Bial grant to collect physiological data (draw blood, track heart rate, etc.) from several WCRMs during 2013 and 2014 with the assistance of a physician. In addition, I am currently (January, 2013) awaiting news regarding funding for a functional magnetic resonance imaging (fMRI) study that may shed light on the survival vs. psi debate. The articles describing the results of those studies will be posted at www.windbridge.org once they are published (with permission from the journal editors).

Psychology

In Step 2 of the eight-step WCRM screening and testing procedure, the prospective WCRM completes several personality and psychological questionnaires, one of which is the Myers-Briggs Type Indicator (MBTI). The MBTI is a popular personality measure based on Carl Jung's theories (you may have taken it yourself). A person's responses to the questionnaire items reflect her preferences in four dichotomies: Extraversion/Introversion (E/I), Sensing/Intuition (S/N), Thinking/Feeling (T/F), and Judging/Perceiving (J/P). The results categorize the test-taker as one of 16 types: ESTJ, INFP, ENTP, etc. No type is 'better' than another and each has its strengths and challenges. (I scored as an ENTJ.)

In the MBTI results from prospective and certified WCRMs I have analyzed, 83% of the mediums were categorized as both N (Intuition) and F (Feeling). According to the makers of the MBTI, only 16% of the general US population is both N and F (I'll call them 'NFs' here) compared to the 83% of research mediums who are NFs. For people whose answer sheets are mailed in to be scored by computers at MBTI headquarters, data is also collected about the test-takers' occupations. The professional groups with the highest proportion of NFs in their ranks are clergy (55% NFs) and art, drama, and music teachers (54% NFs). The professions with the fewest number of NFs are police and detectives with a measly 4%.

This made me think that this may be why mediums and psychics are not always welcomed with open arms (to say the least) by law enforcement officials (LEOs). The two groups simply interact psychologically with the outside world in entirely different ways. It's probably difficult for some LEOs to even conceive of how non-sensory information could be accessed, interpreted, and trusted as valid. Differences in personality types, however, don't seem like reason enough to dismiss a potentially helpful phenomenon in its entirety. This brings us to the third arm of mediumship research at the Windbridge Institute: Application.

❧ 6 ❧

Application Research Program

*When you are born, you cry,
and the world rejoices.
When you die, you rejoice,
and the world cries.*
–Tibetan Buddhist saying

'Applied Research' is right in our name: The Windbridge Institute for Applied Research in Human Potential. The main purpose of applied research is to discover socially useful applications for the phenomena, processes, products, etc., under investigation. This is in contrast to basic research which is simply interested in acquiring knowledge for knowledge's sake.

We perform both basic and applied research at the Windbridge Institute, but when it comes right down to it, we are more interested in how the results of our research can help people than anything else. Who cares how it works? How can we use it? Our mission statement captures this focus:

Julie Beischel, PhD

> The Windbridge Institute for Applied Research in Human Potential is concerned with asking, "What can we do with the potential that exists within our bodies, minds, and spirits?" Can we heal each other? Ourselves? Can we affect events and physical reality with our thoughts? Are we connected to each other? To the planet? Can we communicate with our loved ones who have passed? The Windbridge Institute investigates the capabilities of our bodies, minds, and spirits and attempts to determine how that information can best serve all living things.

Regarding mediumship research specifically, there are two main areas of applied research we are interested in at this time: (a) the contribution of mediumistic information to criminal, missing persons, and other law enforcement cases (that is, forensic mediumship) and (b) the potential therapeutic application of mediumship readings in the treatment of grief.

Law Enforcement

When all sensory means for gathering information about criminal and missing persons' cases have been exhausted, it seems logical that we might want to try some non-sensory methods. Although, using limited time and resources to chase down every psychically-acquired lead doesn't seem like a smart move either. Research needs to be performed in order to assess, for example,

> which types of psychic or mediumistic information about a case (names, locations, physical descriptions, etc.) are accurate most often,
>
> what kinds of crimes impart the most reliable information and which kinds (if any) should be left exclusively to traditional investigation techniques,
>
> what is the length of time after a crime has been reported that is most conducive to accurate and reliable information,
>
> if it's better to use psychics/mediums familiar with the location,
>
> how much information should be provided to the psychic/medium, etc.

In addition, research needs to be performed regarding how to best optimize the acquisition of the psychic information. For example, are there certain physical environments (like the psychic's own workspace), law enforcement personnel (like members of departments who score more toward the NF range of the MBTI), and methods (like open-ended questions) that better aid in the psychic's ability to acquire meaningful information?

In addition to the acquisition of information about cases under investigation, there may actually be a point in the future in which homicide victims can actually testify against their perpetrators. This is obviously quite speculative and a tremendous amount of research needs to take place before the testimony of the dead can be taken seriously, but I wanted to throw it into the 'what if?' category.

At the Windbridge Institute, though we are interested in all of these aspects of forensic mediumship, we do not—at this time—have the resources or personnel to systematically investigate them. Many of the WCRMs on our team have experience working with LEOs (on and off the record) and would be interested in pursuing this line of research, but there just hasn't yet been the time or researchers necessary to address this important applied research direction.

If you are interested in this topic, you may want to look into the work of Find Me— "an organization of talented psychics, law enforcement officers, and professional search and rescue volunteers from all over the world working with law enforcement and families to find missing loved ones and solve homicides"—and/or the experiences of WCRM Debra Martin who provided information in a reading that landed a man in jail for a murder that no one initially even knew happened (see the appendix for a link to the news story).

Grief

A lot is known about grief. For example, we know that a large majority of people get past a loss on their own but that significant mental and physical distress can result from unresolved grief. It is those latter, more severe cases that we hear about the most and that tend to incorrectly reflect what people can expect from their own grief reactions. The facts that (a) there is no standard, 'normal' grief reaction (though there are common experiences) and (b) there are no phases, tasks, or stages that everyone must go through are gaining more and more purchase with professionals and in the literature (hopefully, the public and popular culture will soon follow and we can stop referring to the five or seven or nine stages as somehow real). Unfortunately, the prevalence of extreme stories and references to stages of grief may cause some people to think that they are somehow grieving incorrectly. After hearing the skewed or incorrect information about grief, they may ask themselves:

> I'm starting to feel better. Did I not love this person enough?

> I never felt angry. Am I doing it wrong?

> I can still laugh. Am I a bad person?

The answer to each of those questions, of course, is No. To learn more about the actualities of grief, I recommend *The Truth About Grief: The Myth of Its Five Stages and the New Science of Loss* by Ruth Davis Konigsberg and *The Other Side of Sadness: What the New Science of Bereavement Tells Us About Life After Loss* by George Bonanno.

Another thing we know about grief is that there isn't much we can currently offer the people who are having trouble getting through it. In their 2008 paper "The Effectiveness of Psychotherapeutic Interventions for Bereaved Persons: A Comprehensive Quantitative Review," Currier, Neimeyer, and Berman, researchers at the University of Memphis, reported the results from a meta-analysis (that is, an examination of the combined data from a bunch of studies as if they were just one) of 61 studies about grief and psychotherapy. Their findings revealed a "discouraging picture for bereavement interventions" which they found added "little to no benefit beyond the participants' existing resources and the passage of time." That is, they concluded that traditional psychotherapy provides little to no effect for relieving grief beyond the healing that was going to happen anyway. Pharmaceutical interventions are also relatively useless in treating grief since anti-depressants can take weeks to months to take effect and grief is different than depression anyway.

Research has shown, however, that both induced and spontaneous experiences of after-death communications (ADCs) can dramatically reduce grief. Induced ADCs can involve (a) a system utilizing eye-movement desensitization and reprocessing (EMDR) techniques developed by Allan L. Botkin (described in his book *Induced After-Death Communication: A New Therapy for Healing Grief and Trauma*) and (b) experiences induced by a mirror gazing procedure developed by Raymond Moody which he called a Psychomanteum and which has been studied for over a decade at Sophia University (previously the Institute of Transpersonal Psychology) in Palo Alto, California. Spontaneous ADCs have been shown to be very common and normal and can include sensing the presence of the deceased; visual, olfactory, tactile, and auditory (voices or sounds) phenomena; conversations; powerful dreams; hearing meaningfully timed songs on the radio or music associated with the deceased; messages from objects; natural phenomena; symbols; and synchronicities.

Spontaneous and induced ADCs can have tremendous impacts on the grieving process and my observations as well as pilot data we collected at the Windbridge Institute suggests similar positive effects after readings with mediums.

Julie Beischel, PhD

Grief and mediums

To best convey the impact that experiencing communication with the other side (even through research) can have on someone's grief, I often use the following quote from a man named Bill: "After the devastating loss of two sons, this work has proven to me that we survive the death of our bodies and has made my life not only bearable but worthwhile again." The gravity of that statement —in more ways than one—is remarkable. Bill quite sincerely found living his life to be simply unbearable. Then, after receiving a mediumship reading, he experienced his life as not only bearable but actually worth living. This night-and-day reaction to a mediumship reading is common.

Another story from a research participant that stuck with me was this: Janice and her husband had been contemplating selling their large home and moving into a smaller space when one of their sons moved back in with them after going through a messy divorce. Unexpectedly, Janice's son died while still living there. Six years later (six years!), Janice had a mediumship reading as part of the testing of a prospective WCRM. The medium—who was blinded to all information about Janice and her family beyond her son's first name—conveyed that the male discarnate was saying that the sitter needed to move, needed to sell a house. During the unblinding phase of the study, Janice told the medium and me the story and how she felt she couldn't leave the house because she worried her son wouldn't be able to find her. "He's not in the house," said the medium, "he's with you. And he'll continue to be with you wherever you go." After hearing that, Janice said that her intention was now to sell her house and move (and hopefully move on).

Those examples show how grief can devastate and literally immobilize people and how communication with the deceased through a medium can drastically change lives. (If a pill could have such a dramatic effect on people, the drug companies would be falling all over themselves trying to bring it to market.) Pilot survey data we collected also seem to support this conclusion.

In that exploratory study, we used an anonymous, on-line survey to ask 83 people to retrospectively rate their levels of grief before and after a reading with a medium. The results indicated that they recollected experiencing meaningful reductions in levels of grief. The average grief score went from slightly above "I felt a somewhat high level of grief" to slightly below "I felt a somewhat low, manageable level of grief." A subset of participants (29) also worked with a mental health professional (MHP) as part of their approach. At the end of the survey, we asked, "Do you have any other comments about your grief that you would like to share with the investigators?" The responses to that question were very enlightening.

People talked about the short- and long-term effects of a mediumship reading on their experience of grief:

> *After the reading I felt tremendously upbeat. This euphoria lasted the whole day. It was very amazing.*

> *When I am approached by my loved one that has passed, I am much more accepting of her presence and look forward to the joy instead of the pain... I wish I had had the reading 16 years ago!*

> *[The medium] helped me manage the grief that has been with me for more than 20 years.*

Participants also commented on the relief they felt knowing that this topic was being researched and validated:

> *I get great comfort from visiting afterlife websites and knowing that people who are much more intelligent than me actually believe in an afterlife.*

> *I have tried so hard to let friends understand how helpful the medium was to me but they think I am wrong, that I shouldn't see a medium.*

...consciousness research is important for me in using mediumship readings as a tool in grief recovery.

Finally, several participants specifically commented on the importance of the combination of the two interventions—mediumship readings and work with a MHP—on their recovery:

I believe the combination of assistance from a MHP and a medium to be of significant value in processing grief and corroborating one's belief in life on the "other side."

I can't begin to express how helpful my readings have been in my healing journey. I know that I personally needed to go through counseling as well. However, the level of healing was accelerated by getting readings.

The medium reached my heart, the social worker my mind.

At the Windbridge Institute, we suggest a scenario for the future in which credentialed mediums work together with licensed mental health professionals to address the grief experiences of the bereaved. The role of the medium, in addition to providing communication with the deceased, would be to assist the sitter in locating and beginning work with a mental health professional. Likewise, the role of the mental health professional would be to assist the client/sitter in integrating the information provided by the medium. We believe this scenario is a necessary and practical solution to the increasing use of mediumship readings by the general public and the limited demonstrable effect of traditional grief therapy. We are currently working on bringing this collaborative future to fruition through controlled research and the development of training materials.

I have designed a clinical trial that uses a standard randomization scheme, waiting list control group, group assignment method, quantitative grief instrument, and statistical analysis to analyze the therapeutic effects of a personal mediumship reading from a credentialed, non-denominational, mental medium. This study also aims to investigate the possible correlation between reading accuracy and changes in grief. Due to what I interpret as bias against mediumship phenomena, a funding organization amenable to supporting this project has not yet been located. So we're taking it to the people. If you believe that this project—which will demonstrate if mediumship readings may be a useful therapeutic tool for the bereaved—is worth pursuing, please join our email list at www.windbridge.org to be notified when we launch a crowd funding initiative. (I have been known to say that I will not rest until health insurance companies cover therapeutic mediumship readings.)

Further research will also be needed in order to establish the answers to secondary research questions including:

Who can benefit most from a mediumship reading and which cases require other therapies?

When is the ideal time after the death to receive the reading?

How can a mediumship intervention be best integrated with work with a MHP?

These and other research questions will need to be addressed before any strong conclusions can be drawn regarding the potential therapeutic benefit of mediumship readings in the treatment of grief.

Other Applications

In addition to forensic mediumship and readings as a treatment for grief, mediumship may have other socially relevant applications that we are interested in examining. For example, the deceased may have access to wisdom or knowledge that could benefit scientific, technological, and social progress.

Additionally, mediumship readings may be useful in end-of-life care and hospice situations. As someone who takes solace in the ever-changing conclusions drawn by scientists from ever-expanding data sets, it will be science that brings me comfort in my last moments in this physical body. Perhaps providing mediumship readings to hospice patients and their families about the continued existence and well-being of deceased loved ones may make the transition easier for everyone. To eliminate fear about the next stage of existence, it might be comforting to hear from someone who is already there. The documentation regarding the spontaneous experiences called 'deathbed visions' in which a dying person experiences communication with the deceased near his own death would imply that the human condition already includes this phenomenon. Including mediumship readings as part of end-of-life care would allow all people this comfort. I am regularly saddened by the fact that we in Western cultures spend so much time, energy, and resources to train pregnant women on what to expect and how to best deliver yet we don't give the dying even a hint of what to expect or how to prepare. Perhaps mediumship readings may be part of a new training system for the dying.

Finally, mediumship readings (and survival research in general) may have applications in allopathic healthcare. Although I didn't go to medical school, I'm pretty sure they cover at some point the fact that all bodies die, that death is the inevitable end to every life. Yet, physicians do everything in their power trying to prevent that unavoidable endpoint. This seems like a lot of wasted effort and resources. There have been many anecdotal reports of mediums communicating with people in coma. Perhaps phenomena such as receiving 'permission' from patients to remove life support may aid in a healthcare shift from prolonging the physical life to improving the quality of life and preparing people for the inevitable.

In conclusion, the practical social applications of mediumistic communication and information are numerous and warrant further serious study. I will volunteer to spearhead that endeavor. Who's with me?

Mechanism Shmechanism

One of the criticisms of mediumship and psi abilities in general is that because we can't define clear mechanisms for them, any laboratory demonstration must be the result of fraud, error, chance, statistical manipulation, etc. Some people don't think things are real until they can explain how they occur. However, there are numerous 'normal' phenomena and we can't really explain how or why they happen but we all agree that they exist and are potentially worthy of study.

Some of these are simple things we all have experience with like yawning, dreaming, and blushing and some are diseases and conditions we have at least heard of like multiple sclerosis, lupus, rheumatoid arthritis, Parkinson's disease, eczema, psoriasis, glaucoma, fibromyalgia, and any disease with "idiopathic" in its title.

Because I was trained as a pharmacologist, the unexplained things that come to my mind are the many drugs on the market that work through mechanisms we don't fully understand. These include Botox and Fosamax; aspirin for most of its century of use (though now we know how it works); certain drugs that treat Parkinson's (pramipexole), cancer (procarbazine, targretin), tuberculosis (ethambutol), malaria (halofantrine); and epilepsy (levetiracetam); the antibiotics clofazimine and pentamidine; many psychotropic drugs (for example, lithium); and the general anesthetics that keep patients unconscious during surgery.

In a blog I wrote on this topic, I pointed out that if skeptics need to have surgery, they must then forego the general anesthesia since the doctors cannot define the precise mechanisms of action of those compounds. Those skeptics must be forced to conclude that any previous loss of consciousness demonstrated in other patients when exposed to these drugs was surely due to error, fraud, chance, or statistical manipulation.

Obviously, this is faulty logic. Just because we can't explain how something works doesn't mean we can't employ it as a useful element in our daily lives. At the Windbridge Institute, we care less about how mediumship works than how it can help people.

❦ 7 ❧

So What?

*I believe there is no source of deception
in the investigation of nature which can
compare with a fixed belief that certain kinds
of phenomena are impossible.*
–William James

So what the heck (that's not swearing, is it?) are we supposed to do with all this information?

Obviously, evidence for life after death opens up existential and philosophical questions that I am not qualified to address, but I will now conjecture wildly nonetheless. It seems to me that survival of consciousness research findings demonstrate that there is more to us than the physical matter that makes up our bodies. It would seem that consciousness goes on to exist even without the body. I don't, however, think this principle necessitates a grander spiritual reality. It may simply be a characteristic of consciousness: Consciousness was associated with a body and now it is not. Death need not be any more spiritually transformative than incarnation or birth; they are both tremendous shocks, I can only imagine, but perhaps they are just natural processes that don't necessitate a higher power, a grander meaning, or a deeper causation.

It does, though, seem that connections we make in this physical life tend to extend beyond the physical. That makes sense. Anyone who has experienced a loved one in a physically distant location knows that our connections are not bound by space or time, though they are linked by them, and, therefore, those connections cannot be severed by death. This reality does, in turn, have farther-reaching implications. What about connections to animals? To strangers? To ecosystems? To the future? Obviously, those are questions for another day (when your coffee isn't getting cold). We'll have to pick this up if we meet again in the future.

The Afterlife

The conspicuous question still looming that I have not addressed is a big one: What is the afterlife like? Because the answers to this question are unverifiable (I can't take a camera there and document it), we haven't tackled this issue specifically in our research. For a discussion of the common elements of the afterlife described in channeled material, see *Handbook to the Afterlife* by Pamela Rae Heath and Jon Klimo. It seems that we continue to grow, mature, and heal. And I was quite pleased to hear that our education continues in the afterlife. Yay! Learning! Learning!

The Future

To address where we go from here, let's review what we know to date:

> There are many people called mediums who naturally and normally experience regular communication with deceased people.
>
> Certain mediums are able to report accurate and specific information about the deceased under controlled conditions that account for all normal, sensory methods for acquiring that information.
>
> Though there are several competing non-sensory theories for the source of mediums' information (mainly, survival and psi), communication with the deceased seems, at this time and based on the research I have reviewed here, to be the most parsimonious explanation.
>
> There seem to be unique phenomenological, physiological, and psychological elements characteristic of the mediumship phenomenon.
>
> Several practical social applications of mediumship readings most likely exist.
>
> More research is needed on all of these as well as other fronts.

I think I have made my case for why further research is necessary within the Information, Operation, and Application research programs I discussed, but there are even more questions needing to be tackled. These include, but are not limited to:

Can mediumship be taught, learned, and developed? If so:

Who are the best candidates for learning?

What is the best method for teaching?

Is there a genetic component to mediumship ability?

Can mediums communicate with angels and spirit guides? With extraterrestrial beings? With past lives? With consciousnesses not yet born?

Can mediums provide useful medical information including diagnostic and treatment input?

What are the anthropological, psychological, and methodological differences in the mediumship phenomenon in different cultures?

What are the social implications of belief in an afterlife?

In addition, the phenomenon of physical mediumship warrants further study in order to address questions about, for example, the source of documentable physical phenomena: is it the influence of the dead, psychokinesis by the living, or some of each?

The End is Not the End

As you can see, in regard to the mysterious phenomenon of mediumship, we are faced with far more questions than those for which we currently have answers. I will continue pursuing these and other lines of inquiry and I look forward to sharing my future discoveries with you.

Appendix:
Cited Resources and Materials

Note: Links to source materials are included where available.

Chapter 1

Cited:

Radin, D. (2009). *Entangled minds: Extrasensory experiences in a quantum reality*. Simon and Schuster.

McLuhan, R. (2010). Randi's prize: What sceptics say about the paranormal, why they are wrong, and why it matters. Leicester, UK: Matador.

Fontana, D. (2005). Is there an afterlife? A comprehensive overview of the evidence. Blue Ridge Summit, PA: NBN.

Blum, D. (2006). Ghost hunters: William James and the search for scientific proof of life after death. New York: Penguin.

Also suggested:

Tart, C. T. (2009). The end of materialism: How evidence of the paranormal is bringing science and spirit together. New Harbinger Publications.

Books by Larry Dossey, Rupert Sheldrake, and Chris Carter are widely available from major booksellers.

Video: "Science and the taboo of psi" with Dean Radin
https://www.youtube.com/watch?v=qw_O9Qiwqew

Chapter 2

Wendy. (2000). The naked quack: Exposing the many ways phony psychics & mediums cheat you! Chanworth Global Enterprises.

Beischel, J. (2007). Contemporary methods used in laboratory-based mediumship research. *Journal of Parapsychology, 71*, 37-68.
http://windbridge.org/papers/BeischelJP71Methods.pdf

Chapter 3

Webpage: Windbridge Certified Research Mediums (WCRMs)
http://www.windbridge.org/mediums/

Beischel, J. (2007). Contemporary methods used in laboratory-based mediumship research. *Journal of Parapsychology, 71*, 37-68.
http://windbridge.org/papers/BeischelJP71Methods.pdf

Chapter 4

Blog post: "The Two Sally's" http://drjuliebeischel.blogspot.com/2010/05/two-sallys.html

Beischel, J. (2007). Contemporary methods used in laboratory-based mediumship research. *Journal of Parapsychology, 71*, 37-68.
http://windbridge.org/papers/BeischelJP71Methods.pdf

Chapter 5

Hoffman, D. D. (2000). *Visual intelligence: How we create what we see*. WW Norton & Company.

Rock, A. J., & Beischel, J. (2008). Quantitative analysis of mediums' conscious experiences during a discarnate reading versus a control task: A pilot study. *Australian Journal of Parapsychology, 8*, 157-179.
http://windbridge.org/papers/Rock_Beischel_2008.pdf

Rock, A. J., Beischel, J., & Cott, C. C. (2009). Psi vs. survival: A qualitative investigation of mediums' phenomenology comparing psychic readings and ostensible communication with the deceased. *Transpersonal Psychology Review, 13*, 76-89.

Chapter 6

Find Me organization: http://www.findme2.com/

Local news story about Debra Martin: http://www.kpho.com/story/15919725/the-medium-who-solved-a-murder

Konigsberg, R. D. (2011). *The truth about grief: The myth of its five stages and the new science of loss.* New York: Simon & Schuster.

Bonanno, G. A. (2009). *The other side of sadness: What the new science of bereavement tells us about life after loss.* Basic Books.

Currier, J. M., Neimeyer, R. A., & Berman, J. S. (2008). The effectiveness of psychotherapeutic interventions for bereaved persons: A comprehensive quantitative review. *Psychological Bulletin, 134,* 648–661.

Botkin, A., with Hogan, R.C. (2005). *Induced after-death communication: A new therapy for healing grief and trauma.* Hampton Roads Publishing.

Blog post: "The Mechanism of Mediumship" http://drjuliebeischel.blogspot.com/2012/03/mechanism-of-mediumship.html

Chapter 7

Heath, P. R., & Klimo, J. (2011). *Handbook to the Afterlife.* North Atlantic Books.

Acknowledgements

I am infinitely grateful for the love, support, companionship, and constant entertainment provided by my wonderful husband Mark with whom I am deliriously and madly in love (also Moose, but she can't read). I would like to thank all of my colleagues on the Windbridge Institute Scientific Advisory Board, especially Dean Radin for always taking my phone calls and Loyd Auerbach for ordering desserts and then giving them to me. I thank the other researchers and volunteers on the Windbridge Institute team: Adam Rock, Chad Mosher, Shawn Tassone, Michael Biuso, Danielle, Teresa, Ryan, and Bill. And this research would never have been possible without the generous time, energy, and insight volunteered by the past and present Windbridge Certified Research Mediums. Mark and I are both tremendously grateful for all that Bob and Phran Ginsberg and Forever Family Foundation have done for us over the last five years. This research and its dissemination also could not have been possible without the support, expertise, and resources of Peter, Bill, Abi, David, Mike, Karyl, Lance, Jack, Teri, Susan and David, Sami and Paul, Susan and Chuck, and the valuable members of the Windbridge Institute. I'd also like to thank Larry Dossey because he's just a swell guy and I always enjoy our time together. I thank the Bial Foundation for their continued support of our research. I am also grateful for my 'healers' Nancy, Anita, and Gerry and all the family (biological, in-law, and chosen), friends, and neighbors with me on this ride especially Jean, Audrey, Cyn, Janet, Joe, Julia, Katie, Dominique, Chad and James, Katie and Mikal, Gerry and John, Janette and Sean, Nicole and Andrew, and Cheryl and Eric. Finally, many thanks go to Jaclyn Vigil for her editorial assistance.

Julie Beischel, PhD

# Meaningful Messages

Making the Most of Your Mediumship Reading

Julie Beischel, PhD

The Windbridge Institute, LLC
1517 N. Wilmot Rd. #254
Tucson, AZ 85712
http://www.windbridge.org/
info@windbridge.org

Text copyright © 2013 by Julie Beischel, PhD

All Rights Reserved

First e-book edition September 2013

The information in this publication is provided "as is" without warranty of any kind, either express or implied, including but not limited to the implied warranties of merchantability and fitness for a particular purpose. Under no circumstances shall the Windbridge Institute, LLC, nor any party involved in creating, producing, or delivering this publication be liable for any damages whatsoever including direct, indirect, incidental, consequential, loss of business profits, or special damages.

Always consult with your physician or other qualified healthcare provider when seeking treatment options.

Praise for *Meaningful Messages*...

"Great advice for how to prepare for your next mediumship reading. Well-thought-out and easy to read these tips will help you get the most out of your experience."
–Chance Houston, paranormal investigator

"As a medium, I am asking all of my clients to read *Meaningful Messages: Making the Most of Your Mediumship Reading* by Julie Beischel before they get a reading from me or any other medium. The main issues that I need a client to know about are all covered in this short, enjoyable read."
–Dave Campbell, Windbridge Certified Research Medium

Acknowledgements

I am grateful for all of the ongoing support and encouragement and the ear-to-ear smiles I get that originate from my husband, Mark, and our awesome dog, Moose. I am thankful for our friends and neighbors, the members and advisors of the Windbridge Institute, and all of the people who are just generally supportive of what we do. Thanks also go to Jaclyn Vigil for her editorial assistance. Finally, I would like to specifically thank the Windbridge Certified Research Mediums; they have taught me so much about mediumship.

Meaningful Messages Contents

∾181∾ Preface

∾193∾ Tip 1: Be Patient

∾195∾ Tip 2: Choose Wisely

∾198∾ Tip 3: Ask Your Discarnate to Be There

∾199∾ Tip 4: Remember That Each Medium's Process is Different

∾200∾ Tip 5: Provide Just the Right Amount of Information

∾204∾ Tip 6: Remember That Even the Best Mediums (and Discarnates) Aren't Perfect

∾207∾ Tip 7: Remember That You Are Grieving

∾209∾ Tip 8: Document the Information and Review it Later

∾210∾ Tip 9: Don't Create Codes or Ask "Proof" Questions

∾212∾ Tip 10: Note or Request Information About How You Can Interact with Your Discarnate After the Reading

∾214∾ Conclusions

Julie Beischel, PhD

When you are born, you cry,
and the world rejoices.
When you die, you rejoice,
and the world cries.
—Tibetan Buddhist saying

What the caterpillar calls the end
the rest of the world calls a butterfly.
—Lao-tzu

Death is not extinguishing the light;
it is putting out the lamp
because the dawn has come.
—Rabindranath Tagore

Julie Beischel, PhD

Preface

I have been performing scientific research with mediums full-time for 10 years. I received my PhD in Pharmacology and Toxicology with a minor in Microbiology and Immunology from the University of Arizona in 2003 and am currently the Director of Research at the Windbridge Institute for Applied Research in Human Potential, an independent research organization.

I am often asked for advice or suggestions about visiting a medium. I wrote *Meaningful Messages* to provide helpful do's and don'ts for people interested in receiving a reading from a medium. It's a short list of 10 tips to keep in mind as you prepare for, experience, and reflect on a reading from a medium.

Although this is an important and serious topic, I have tried to write about it conversationally (no one likes a lecture) and in my own voice which is often laced with humor. As George Bernard Shaw once said, "Life does not cease to be funny when people die any more than it ceases to be serious when people laugh." I hope you will not find any humor irreverent.

Preparation

Having realistic expectations about a reading with a medium will make the whole process go more smoothly for the (at least) three people involved in a mediumship reading: you, the medium, and your deceased loved one. This can be difficult with all of the inaccurate information about mediums that has been presented in movies and on TV. For example, because mediums are often written or edited to be nearly perfect in the accuracy of their statements, people who expect all mediums to perform at this level will invariably be disappointed. Hopefully, *Meaningful Messages* will provide you with the necessary information required to get a successful reading from a real medium.

Please note that specific tips from the medium performing your reading will always trump what I have written here. Her instructions should carry far more weight than my suggestions.

Let's first just get some vocabulary out of the way. A medium is a person who regularly experiences communication with the deceased. A psychic is a person who regularly experiences information about living people, distant locations or events, and/or times in the future or in the past (that they did not live through originally). Though we are potentially all capable of mediumistic and/or psychic experiences, only those people who have those experiences regularly are called mediums and psychics.

It is often said that all mediums are psychic but not all psychics are mediums. And because 'medium' is a common word with several meanings, sometimes to be clear a medium who communicates with the deceased is called a psychic medium.

For ease in reading this book, I have used some shorthand. So that I am not repeating phrases like 'your departed loved one' and 'the deceased person you lost,' I have chosen to describe these loved ones who have passed with the term we use in research: discarnate.

The word discarnate comes from the roots 'dis' meaning 'not' and 'carn' meaning 'flesh.' A discarnate is someone who is not in the flesh. It is just a general term for a person who is deceased. Though a somewhat clinical term, its use means no offense and I hope that each time I use it, you will 'hear' the respect with which it was intended.

I have also used male pronouns (he and his) to describe discarnates and female pronouns (she and her) for mediums. This, of course, does not mean that the information does not apply to female discarnates or male mediums.

The process during which a medium experiences and conveys information from a discarnate is called a reading. The living person requesting that the medium connect with and report communication from the discarnate (you) is called a sitter or a client.

Although some sitters receive mediumship readings out of pure curiosity, most are hoping to reconnect with their discarnates in order to alleviate their grief. Recently, there has been a sort of grassroots movement of people seeking non-traditional treatment options like mediumship readings to manage their grief over a death when perhaps nothing else has worked. Scientific research has not yet caught up with this widespread practice and investigated the potential therapeutic benefit of mediumship readings for the bereaved. However, I have taken the initial steps. In addition, there is extensive research demonstrating the benefit of similar after-death communication (ADC) experiences on grief.

Spontaneous and Assisted After-death Communication

In contrast to more mainstream therapies, research has demonstrated that non-traditional experiences can positively and dramatically impact peoples' grief after the death of a loved one. And though these spontaneous experiences of communication with the deceased are often called 'extraordinary' or even 'paranormal,' roughly one-third of American adults from all types of socioeconomic and religious backgrounds have had after-death communication experiences (or ADCs).

These spontaneous ADCs seem to be part of the normal grieving process and can include sensing the presence of the deceased; seeing, hearing, or feeling the deceased; scents or odors associated with the deceased; dreams; flickering lights; symbols; synchronicities; and other unexplainable occurrences. Numerous researchers have demonstrated how these experiences significantly impact the bereaved by offering comfort and peace and alleviating anxiety and despair.

Receiving a reading from a medium may serve as an assisted ADC and have similar effects on the bereaved. Anecdotal reports and an informal survey study we performed would also suggest as much.

These assisted ADCs may even have advantages compared to their spontaneous counterparts. For example, a reading may be more preferable to people who may be fearful of unexpected contact from a discarnate. (Surprise!) With the medium facilitating the contact in a controlled environment at a scheduled time, it may be a less stressful experience for a nervous or anxious sitter. A mediumship reading may also be useful for sitters who desire contact but have not yet experienced it. In addition, a medium may serve as collaborator in the ADC without disbelieving, dismissing, or ridiculing the experiences or beliefs of the sitter who may find it difficult to locate others who are as supportive.

Although these are logical arguments for how a mediumship reading might be helpful for the bereaved, there have been no systematic, empirical studies published investigating the effects of a personal reading from a modern-day, secular medium. Although I have designed a randomized clinical trial to address this issue, the study is not yet complete.

Researchers and the mental health community in general do not at this point know for certain if a mediumship reading is helpful, hurtful, or neither for the grieving and, if it is helpful, who might benefit the most and when. Controlled research is required.

Thus, I am not endorsing mediumship readings which may not be helpful or even appropriate for everyone. The decision to receive a reading is not one that should be taken lightly. A reading could have profound effects (negative or positive) so deciding to receive one should be done with significant forethought and intention.

If and when you have decided to receive a reading, the 10 tips presented in *Meaningful Messages* should help you make the most of it.

Things a Medium Might Say

First and foremost, it is important to remember that which discarnates and which items come through are not up to the medium. The mediumship process involves the medium opening up to receive information from your discarnate. What she sees, hears, smells, tastes, and feels is not up to her any more than what you hear when you answer a ringing telephone is up to you. Keep in mind that all she can do is tell you what your discarnate is telling or showing her and provide her interpretations of the perceptions and symbolism.

If you think this process is easy, then you shouldn't need a medium because you should be able to contact the dead all by yourself. I invite the rest of you to use that logic on anyone who disparages your reports of your reading with assumption-laden statements like, "Well, she should've been able to tell you [blank]." If it was so easy, we could all just do it ourselves.

When a medium conveys information during a reading, she may use phrases like "he is showing me" or "I am seeing." Alternatively, the medium may quote the discarnate directly with items like "I really liked the tofu scramble you made on Sunday mornings." Sometimes mediums will use the present tense ("He is...") and sometimes they will use the past tense ("He was..."). The content of the items is much more important than the way the medium phrases them.

There are three major types of information that are reported most often by mediums during readings. The first type is information that helps the sitter identify the discarnate. This can include descriptions of the discarnate when he was living such as his physical appearance (e.g., hair and eye color, height, build, unique scars or birthmarks), personality characteristics, other deceased people or animals with him, and favorite activities, foods, events, places, etc. This information helps the sitter feel confident that the information is coming from the 'right' discarnate.

The second type of information describes events in the sitter's life that have occurred since the death. A medium may say things like "She was at the memorial you put together," "He likes your new haircut," or "He walked you down the aisle." This type of information provides evidence that the discarnates continue to observe and participate in the sitters' lives.

The third type of information is messages specifically for you, the sitter. This can include simple messages like "I love you" to messages that seem to be for the purpose of alleviating guilt or sadness like "There was nothing you could have done to prevent his death" or "I didn't feel any pain." Their messages may also be similar to what they would say to you when they were living. They might chastise or reprimand you for things they think you shouldn't have done or things they think you should have done by now. They might offer advice about your finances, your career, your relationships, or your lifestyle. They might offer encouragement for things you are thinking about or attempting.

The important piece to remember is that you have free will and you shouldn't take the advice of anyone (living or deceased) without deciding what would be best for you. Discarnates are just people without physical bodies. They don't have all the answers to the universe, so their opinions are simply that: opinions. People do not become immediately all-knowing and self-actualized when they die; they are still people with their own individual opinions, beliefs, and idiosyncrasies. You should keep these factors in mind before taking the advice of any medium or discarnate. Mediumship readings should never replace the advice of appropriate physicians, lawyers, counselors, or other professionals.

If and when you decide to have a medium attempt to contact your discarnate, there are several things you can do to optimize the process for the medium, you, and your discarnate so that a successful reading can take place. Each of the following sections describes a different tip or suggestion to that end.

Notes and Further Reading

Beischel, J. (2013). *Among mediums: A scientist's quest for answers*. www.amazon.com/dp/B00B1MZMHM/

Beischel, J., Mosher, C. & Boccuzzi, M. (2014-2015). The possible effects on bereavement of assisted after-death communication during readings with psychic mediums: A continuing bonds perspective. *Omega: Journal of Death and Dying, 70*(2), 169-194. doi: 10.2190/OM.70.2.b

Bonanno, G. A. (2009). *The other side of sadness: What the new science of bereavement tells us about life after loss*. New York: Basic Books.

Conant, R. D. (1996). Memories of the death and life of a spouse: The role of images and sense of presence in grief. In Klass, D., Silverman, P. R., & Nickman, S. L. (Eds.), *Continuing bonds: New understandings of grief* (pp. 179–196). Washington, DC: Taylor & Francis.

Daggett, L. M. (2005). Continued encounters: The experience of after-death communication. *Journal of Holistic Nursing, 23*, 191–207.

Dannenbaum, S. M., & Kinnier, R. T. (2009). Imaginal relationships with the dead: Applications for psychotherapy. *Journal of Humanistic Psychology, 49*, 100–113.

Drewry, M. D. J. (2003). Purported after-death communication and its role in the recovery of bereaved individuals: A phenomenological study. *Proceedings of the Annual Conference of the Academy of Religion and Psychical Research*, 74–87.

Haraldsson, E. (1988-89). Survey of claimed encounters with the dead. *Omega: Journal of Death and Dying*, *19*, 103–113.

Houck, J. A. (2005). The universal, multiple, and exclusive experiences of after-death communication. *Journal of Near-Death Studies*, *24*, 117–127.

Klugman, C. M. (2006). Dead men talking: Evidence of post death contact and continuing bonds. *Omega: Journal of Death and Dying*, *53*, 249–262.

Konigsberg, R. D. (2011). *The truth about grief: The myth of its five stages and the new science of loss*. New York: Simon & Schuster.

LaGrand, L. E. (2005). The nature and therapeutic implications of the extraordinary experiences of the bereaved. *Journal of Near-Death Studies*, *24*, 3–20.

Normand, C. L., Silverman, P. R., & Nickman, S. L. (1996). Bereaved children's changing relationships with the deceased. In Klass, D., Silverman, P. R., & Nickman, S. L. (Eds.), *Continuing bonds: New understandings of grief* (pp. 87–111). Washington, DC: Taylor & Francis.

Nowatzki, N. R. & Grant Kalischuk, R. (2009). Post-death encounters: Grieving, mourning, and healing. *Omega: Journal of Death and Dying, 59*, 91–111.

Parker, J. S. (2005). Extraordinary experiences of the bereaved and adaptive outcomes of grief. *Omega: Journal of Death and Dying, 51*, 257–283.

Sanger, M. (2009). When clients sense the presence of loved ones who have died. *Omega: Journal of Death and Dying, 59*, 69–89.

Sormanti, M. & August, J. (1997). Parental bereavement: Spiritual connections with deceased children. *American Journal of Orthopsychiatry, 67*, 460–469.

Tip 1:
Be Patient

Do not rush into a reading before you and your discarnate are ready. Remember that your discarnate may need to adjust to the death just like you needed to. If your discarnate's death was a shock to you or difficult for you to accept, it might be helpful to consider the shock it was to him.

We don't know exactly what happens after someone dies. Maybe there is the equivalent of welcome packets or orientation classes. Maybe there is paperwork to fill out. We just don't know. Allowing your discarnate to "get settled" before you request communication is a gift that you can give that may make the transition easier for him.

If he is learning the ropes, it might be difficult to become acclimated if he is aware of your calling for him. In addition, figuring out how to communicate with a medium now that one is deceased might take some time and practice.

It seems that your "wanting it too badly" may even prevent a medium from being able to connect with your discarnate. There are also numerous reports of successful readings happening several decades after the death occurred, so know that you are not up against any kind of deadline.

Simply "putting it out there" that you are willing to wait and requesting that your discarnate send you a sign letting you know when he is ready to communicate may make the situation less stressful for everyone involved.

In our research, we require that the death has occurred at least one year ago in order for sitters to participate in research readings. I have heard of mediums who also set this guideline for their potential clients. This seems to be a good amount of time to wait to allow your personal resources (e.g., friends, family, the passage of time) to impact the severity of your grief.

(For more about the transition/dying process, see *Handbook to the Afterlife* by Pamela Rae Heath and Jon Klimo.)

Tip 2:
Choose Wisely

Although there are unscrupulous characters in any profession, mediumship unfortunately has a long history of frauds and con artists. These phony mediums often use a process called cold reading in which they use visual and auditory clues from the sitter to steer the reading and make it appear that they are reporting information they couldn't possibly know. This can involve the medium noting simple things like a widening or narrowing of the eyes, subtle changes in facial expressions, or slight gasps or other vocalizations as they fish for information.

Because the grieving are vulnerable and often would do almost anything to hear from their discarnates, they are easy prey for charlatans wanting to make a buck. And without legal regulations, anyone can call herself a medium, hang out a shingle, and charge the grieving for readings.

Therefore, it is essential that you choose a medium with care. For example, request information about the qualifications or experience of any medium you are considering. This is especially important if she uses terms like "certified," "tested," or "trained." Ask the medium to describe the certification, testing, or training. Often, these terms are bestowed on mediums who simply pay a fee to be able to claim such experience or be listed on a website. The best screening procedures are the ones that are the most transparent; the procedure used by the certifying organization should be easily obtainable.

Furthermore, even claimant mediums themselves may not recognize when they are using cold reading, general information, or other normal psychological techniques to create accurate readings. They may not be able to report accurate information without a sitter steering them but are not aware of that fact. These claimant mediums are not necessarily intentional frauds (though they might be), but they also are not necessarily able to provide true mediumship readings.

To find a medium, you may wish to get recommendations from people or organizations that you trust. To select a medium from the list suggested to you, we recommend you start by looking at the mediums' websites. One will most likely "speak to you" (pun intended). You can even ask your discarnate to help you choose.

It is important to note that a medium does not need to be in your city, state, or country in order for you to receive a good reading. Phone readings can actually be better than in-person readings for several reasons:

You can be sure the information the medium is reporting does not come from what she can glean from seeing you.

The mediumship process seems to be a 'right-brained' or intuitive activity and if she can see you, her 'left brain' or analytical side may get in the way by trying to make assessments about you from your appearance, age, etc. This is one reason why authentic mediums may close their eyes during in-person readings.

Any visible expressions of grief you may exhibit will not distract her. If the medium you are seeing is an empathetic one (and that is not a given), the pain you are demonstrating can divert her attention.

She can perform readings in an environment that is comfortable for her and, thus, conducive to her process. With phone readings, the medium is not required to, say, vacuum, put on her shoes, or prepare for a guest in other ways.

On the phone, a medium can just concentrate on receiving the communication from your discarnate who will be able to find her either way.

Finally, it is important to remember that an expensive or famous medium does not necessarily provide better readings than one who is more affordable or less well-known. I was at a conference with some mediums once and during a break, one visited a local psychic fair and got a reading for $20. She was quite impressed. More money does not equal more information.

Tip 3:
Ask Your Discarnate to Be There

Prior to the reading, let your discarnate know that you would like to hear from him and you hope he will participate. You can do this when you schedule the reading or just before the reading takes place. You can make the request in your head like a wish or a prayer.

This step conveys your intention and your willingness to hear from your discarnate in this way. It makes it clear that you are a willing participant in the process.

Requesting that your discarnate participate may also give your permission to the medium to communicate with your discarnate since often the medium receives information as she prepares before a reading starts.

Tip 4:
Remember That Each Medium's Process is Different

Forget about what you've seen on TV or how other mediums' readings have gone. Each medium knows the process that works best for her and it is best if you just go with it.

Some mediums begin each reading with a meditation, prayer, or the setting of an intention. Often this takes place before the reading, but if you are present for it, accept that it is a necessary part of her process. Your worldview does not need to match the medium's. You should, however, be respectful of her beliefs and practices.

Often a medium will begin by describing her process to you so that you know what to expect. She may, at that time, ask if you have any questions about the process.

Tip 5:
Provide Just the Right Amount of Information

During the reading, you may want to both (1) be aware of the information that the medium could guess just by talking to or seeing you and (2) provide as little additional information as possible.

As I mentioned in Tip 2, though fraudulent mediums may use the information they get from you to fake accurate readings, even mediums with the best of intentions might not realize that they have subconsciously used clues you provide during a reading. It is important to recognize the potential for these clues so that you do not assign more meaning to an item than is warranted.

For example, if you have a British accent, you may not want to take a medium's reports of images of Big Ben as evidence that she is definitely communicating with your deceased father (unless perhaps his name was Ben and you hadn't told her that). If you visit a medium with your three sisters and you each have names that start with the letter P, you probably should not be impressed if she reports that your deceased sister's name also started with a P.

However, please also take into account that the mediumship state appears to be a somewhat altered state of consciousness and the medium may not remember information she acquired outside of that state or even earlier in the reading. Basically, don't assume fraud or cold reading at every turn, but use common sense and don't be overwhelmingly convinced by every 'hit.'

Furthermore, keep in mind that a mediumship reading is a unique situation in which you and another person are focused specifically on your discarnate(s) for an extended amount of time. This is most likely a rare condition in your everyday life and becomes even rarer as more time passes after the death and it stops being socially 'acceptable' for you to talk about your discarnate at length. Be aware that you will have a tendency during a mediumship reading to accentuate this scenario by telling the medium all about this person that you loved and lost. (Believe me. Though you might think you'll be able to control yourself, the information may just come pouring out of you. Just keep that potential inclination in mind.)

However, not only does telling the medium everything about your discarnate in one breath not help her, but it actually impedes her ability to communicate with him. I think most authentic mediums would prefer sitters didn't say anything at all.

In addition, you "steal" any information you provide to the medium from her. Just think how much more evidential the reading would be if the medium told you your brother's nickname or your mom's favorite food rather than you telling her.

It does seem that it is beneficial to the mediumship process to initially provide the medium with some information about the discarnate such as his first name and/or your relationship. In our research readings, we provide the medium with the first name of the discarnate which appears to help the medium focus and may aid the discarnate in focusing on the medium.

During the reading, you can try to limit yourself to answering the medium's questions—if she has any—with responses like "Yes, that makes sense," "No, that's incorrect," "I'm not sure," "Sort of," "Maybe," etc. Often a medium will simply ask, "Does that make sense to you?" and you should try to use only responses like those listed and not give any further information (which, again, you are really going to want to). Try to let the medium and your discarnate figure it out without your assistance.

Still, be careful with your "No"s. Don't say "No" when you don't know, but say "No" when you're sure. For the reasons stated below in Tip 6, it may not be appropriate for you to adamantly dismiss information reported during a reading that doesn't immediately make sense to you. Unless you are positive about something, it is best to say you're not sure or you don't know or even "I don't think so."

However, if you're quite sure your daughter's middle name wasn't Natalie, then say so. It doesn't help the medium or the process any to try and make everything she says fit. Because a lot of interpretation about the medium's perceptions occurs during a reading, if you provide feedback that the interpretation might be wrong, then she can request additional information from the discarnate to help her interpret the symbol or perception accurately.

Tip 6:
Remember That Even the Best Mediums (and Discarnates) Aren't Perfect

I think it bears repeating here that mediums you see on TV are often heavily edited to remove the items they report that didn't make sense to the sitters. This makes it look like everything they say is a hit. This is not realistic and expecting a real-life medium to perform at that level will hinder the reading and leave you disappointed.

Though it would be nice if discarnates could call mediums on the phone and clearly speak the messages they want to convey, that's not how it works. There are layers of symbolism and interpretation throughout the process.

A medium may receive information through all of her mental senses. She may see images in her mind's eye; hear words, music, or other sounds in her mind's ear; smell perfume, cigars, hospitals, etc. (in her mind's nose?); taste the discarnate's favorite foods; and feel the discarnate's ailments or cause of death in her body. She then has to interpret and translate her perceptions to make sense to you. This is not always easy.

During experiments, we ask our research mediums to "say what you see" and then provide an interpretation of the meaning. Sometimes, the medium may be corrected by the discarnate if her interpretation is incorrect.

In addition, we don't know what it's like to be dead and trying to communicate with the living. Just like it's detrimental to the process to make assumptions about how mediumship works, it isn't helpful to make assumptions about the capabilities of the deceased either. There may be some types or pieces of information that are difficult to convey.

Furthermore, since a mediumship reading involves people communicating with people, not everyone is going to get along every time. I'm sure you've met people with whom you just didn't connect or who you just didn't like. Perhaps not all mediums get along with all discarnates. Because the living regularly have trouble seeing eye to eye, it isn't surprising that a discarnate may have trouble getting his point across to a medium or that a medium may find a particular discarnate difficult to understand. It might be that your discarnate just doesn't like the medium you chose (or vice versa) and is, therefore, hesitant to communicate or refuses all together. This is why it is important to be 'inspired' by your discarnate to choose a medium he likes.

We have also witnessed readings in which discarnates seem to refuse to talk to mediums who have repeatedly provided accurate and specific information for other sitters. This can mean that the discarnate feels the sitter isn't ready to hear from him even if the sitter adamantly claims otherwise. Often, these sitters are the parents of children who died suddenly or violently. Keep in mind that if your discarnate doesn't come through, it might be a sign that you have some additional healing to do before you have a reading. (It, of course, may also mean that the medium isn't very good.)

Finally, it seems that mediums often experience correct and incorrect information in the same way. This makes it difficult for them to know which items you will consider accurate and which you won't. It's not helpful (and sometimes not possible) for a medium to edit herself so she usually just reports everything she is experiencing and trusts that you will get the messages intended for you.

Tip 7:
Remember That You Are Grieving

The process of grief is also a sort of altered state, so note that all of your perceptions, needs, recollections, etc., are colored by the grief. For example, the bereaved may have a tendency to remember readings as being more accurate than they actually were. Grieving sitters may remember the hits and forget about the misses.

Do what you can to protect and take care of yourself while in the unique bereavement condition. Don't put too much pressure on a reading to prove to you that relationships can continue even after a death. Just know it.

Consider employing the assistance of a licensed mental health professional (MHP) as part of this process. It may be beneficial to find an open-minded counselor to help you decide when you are ready to hear from your discarnate and to help you integrate the information from the reading afterward.

In 2012, the American Center for the Integration of Spiritually Transformative Experiences (ACISTE, pronounced 'assist') began training and certifying

MHPs, spiritual directors, pastoral counselors, and other professionals in private practice as well as peers to be knowledgeable about spiritual experiences so that experiencers have a place to turn for support. ACISTE posts a Support Directory on their website listing people who have been certified as well as support groups listed by state.

You can also call around to the counselors in your area, saying something like:

"I am considering seeing a medium for a reading and am calling to see if you can help me prepare for that. Are you open to that practice as a part of the healing process? Is it something that you have any training in or experience with?"

A good, objective friend might also be able to assist you in deciding how a reading might fit in with where you are in the grieving process.

Tip 8:
Document the Information and Review it Later

There will be a lot happening for you during a mediumship reading. You may be generally nervous or anxious about the process and you may experience a number of different emotions during the reading based on the content.

Thus, you may later remember things reported in the reading that you inexplicably forgot at the time. This happens surprisingly more often than one might expect. We have witnessed sitters forget their middle names, the existence of a family member with a certain name, the fact that someone served in the military, etc. It's best to review the information again once you are in a clearer state of mind.

In addition, it is common for a reading to include information unknown to you but that can be verified by others or information that is not yet true but can happen in the future. For these reasons, it is important to have a record of the information stated during the reading so that you can return to it with new information or at a later date. Some mediums will provide you with a recording of your reading. If not, ask her if you can record it or take notes.

Tip 9:
Don't Create Codes or Ask "Proof" Questions

It might seem logical to ask the medium a question to which only you and the discarnate know the answer. This might seem like a good test to prove that she's for real and talking to the right discarnate. However, this logic is quite flawed.

First, a medium's accurate answer to a 'test' question can easily be attributed to other psychic (or psi) abilities and not proof of communication with your discarnate. That is, she could be reading your mind or someone else's, remote viewing the envelope where you sealed the answer, using precognition to access the future, etc. If you or anyone else knows (or will know) the answer to the test question, the medium could just be getting it from you. Thus, it's not an appropriate test to prove she is communicating with the deceased. (For decades, researchers have been struggling with this 'Survival vs. Psi' debate about the source of the information mediums report and there is no type of information that can be attributed to one and not the other.)

Second, it might not be possible for discarnates to convey complex, numerical, or otherwise difficult answers or codes. Maybe they have forgotten some things they knew when they were alive. Maybe they just don't feel like it. These unknowns can make quizzing the dead a disappointing endeavor.

Rest assured that your discarnate will provide information during the reading that will be uniquely identifying of him without the need for you to request items like the combination to a lock or the color of the dress you wore to prom.

Tip 10:
Note or Request Information About How You Can Interact with Your Discarnate After the Reading

We've heard mediums say that their main goal is to put themselves out of business, meaning that the purpose of a reading is to connect you with your discarnate, help you redefine your relationship, and share with you ways you can continue communicating with him without the assistance of the medium. Often, information about the signs or symbols your discarnate may use to interact with you will be revealed during a reading. If it isn't, specifically ask your discarnate to tell you how he has been interacting with you in the past or how he plans to in the future or even choose a sign yourself that the two of you can use.

It is important that you are able to continue connecting with your discarnate after the reading is over or it may seem like you have lost him all over again or that you need the medium to keep the relationship going. This can result in a dependency on mediums.

It is important that you view the mediumship reading as (1) a reminder that you are still connected to your discarnate and (2) a tool for learning how to communicate with your discarnate yourself.

Conclusions

A mediumship reading is a unique opportunity and having the information detailed in the tips listed here will hopefully optimize the experience for you, the medium, and your discarnate. With these 10 tips in mind, I hope you will be able to successfully prepare for, participate in, and reflect on a mediumship reading with ease.

INVESTIGATING MEDIUMS

Julie Beischel, PhD

From the Mouths of Mediums

Conversations with Windbridge Certified Research Mediums

Volume 1: Experiencing Communication

Contributors:

Ankhasha Amenti, Traci Bray, Dave Campbell, Carrie D. Cox, Joanne Gerber, Daria Justyn, Nancy Marlowe, Sarah McLean, Tracy Lee Nash, Troy Parkinson, Ginger Quinlan, Eliza Rey, Kim Russo

Edited By Julie Beischel, PhD

The Windbridge Institute, LLC
1517 N. Wilmot Rd. #254
Tucson, AZ 85712
http://www.windbridge.org/
info@windbridge.org

Text copyright © 2014 Julie Beischel, PhD

All Rights Reserved

First e-book edition July 2014

The information in this publication is provided "as is" without warranty of any kind, either express or implied, including but not limited to the implied warranties of merchantability and fitness for a particular purpose. Under no circumstances shall the Windbridge Institute, LLC, nor any party involved in creating, producing, or delivering this publication be liable for any damages whatsoever including direct, indirect, incidental, consequential, loss of business profits, or special damages.

Always consult with your physician or other qualified healthcare provider when seeking treatment options.

*To Susy Smith
who continues to inspire
mediumship research
at the Windbridge Institute*

Praise for Vol. 1 of *From the Mouths of Mediums...*

"Vol. 1 of *From the Mouths of Mediums* is destined to become an indispensable toolkit for anyone planning on visiting a medium or for anyone wanting to become one... Julie Beischel delivers a metaphysical knockout without pulling any punches. Can't wait for the next volumes to materialize."
–Marcel Cairo, medium and researcher

"Highly informative and deeply provocative for both the experienced and inexperienced, this inspirational and comforting book demonstrates that mediums are just people, too, and not the oft-portrayed unrealistic versions seen in various media."
–August Goforth, psychotherapist, medium, and co-author of *The Risen: Dialogues of Love, Grief, & Survival Beyond Death*

"Vol. 1 of *From the Mouths of Mediums* offers a fascinating insight into the processes and experiences of spirit mediums. Far from the shadowy figures demonized by outspoken skeptics, the Windbridge Institute-approved mediums interviewed for this book are shown to be caring, feeling human beings with as much curiosity about what they do as the scientists that are currently studying them. Recommended reading for anyone interested in the phenomenon of mediumship!"
–Greg Taylor, author of *Stop Worrying! There Probably Is an Afterlife*

"This is a great collection of subjective understandings by mediums of their own work and the ways non-mediums can receive after-death communications (ADCs) on their own. It shows how individual mediums have their own 'dictionaries' of what means what, in addition to the general culture of mediumship."
–Charles F. Emmons, sociologist at Gettysburg College, author (with Penelope Emmons) of *Guided by Spirit: A Journey into the Mind of the Medium* and *Science and Spirit: Exploring the Limits of Consciousness*

"Dr. Beischel gets inside the very souls of 13 living mediums and lets us see the world through their eyes. We see how they can tell they are making contact with the deceased. We learn the part that we, the living, must play if we are to communicate with our loved ones. And we get more than a glimpse of the world they live in. Finally, we see what we must do if we aspire to mediumship ourselves."
–Stafford Betty, author of *The Afterlife Unveiled* and *Heaven and Hell Unveiled*

"By focusing on the experiences of mediums, described in their own words, *From the Mouths of Mediums* provides fascinating insight into the underlying subjective processes of mediumistic readings."
–Jack Hunter, co-editor of *Talking with the Spirits: Ethnographies from Between the Worlds*

Julie Beischel, PhD

Acknowledgements

I am so grateful for all of the Windbridge Institute members and other supporters who recognize the importance of independent research for examining topics like mediumship. The loving support provided by my wonderful husband and research partner, Mark, and our awesome 14-year-old, 65-pound dog, Moose, always surrounds me and for that I am tremendously thankful. I have told her this privately, but I would also like to publicly thank Sue R. for the way she allowed me to witness up close and personal the natural process of her grief as it threatened to consume her and then morphed into something tolerable and then even fortifying, and I thank the universe for bringing her into my life. Appreciation also goes to Jaclyn Vigil and Katie Mullaly for their editorial assistance. Finally, I would like to specifically and enthusiastically thank the Windbridge Certified Research Mediums who participated in this volume: What you have to say is important in demystifying normal experiences like yours and I'm honored that you allowed me to share your words.

Julie Beischel, PhD

From the Mouths of Mediums
Contents

- ~ 227 ~ Editor's Introduction to the Series
- ~ 237 ~ Editor's Introduction to Volume 1
- ~ 241 ~ Part 1: Experiences of Communication
- ~ 243 ~ Chapter 1: The Reading Ritual
- ~ 249 ~ Chapter 2: Receiving Communication
- ~ 271 ~ Chapter 3: Specific Cases
- ~ 281 ~ Part 2: Advice for Communication
- ~ 283 ~ Chapter 4: General Advice
- ~ 286 ~ Chapter 5: Types of Communication
- ~ 293 ~ Chapter 6: Specific Suggestions
- ~ 301 ~ Part 3: Absence of Communication
- ~ 303 ~ Chapter 7: Discarnate Communicator Factors
- ~ 307 ~ Chapter 8: Living Experiencer Factors
- ~ 318 ~ Chapter 9: Still Not Receiving Communication?
- ~ 320 ~ *From the Mouths of Mediums Vol. 1* Contributors

Julie Beischel, PhD

Editor's Introduction to the Series

The dead are invisible, not absent.
—Saint Augustine

I serve as Director of Research at the Windbridge Institute and my main research focus includes studies of mediums: individuals who experience regular communication with the deceased. Psychics, on the other hand, regularly experience acquiring information from living people (telepathy), about distant locations (clairvoyance), and/or times in the future (precognition) or in the past (that they did not personally experience; retrocognition). Conventional wisdom states that all mediums are also psychics but not all psychics are mediums. Additionally, it is quite common for people who are not mediums or psychics to have periodic mediumistic or psychic experiences.

My research with mediums examines the accuracy and specificity of the information they report; their unique experiences, psychology, and physiology; and the practical social applications of mediumship readings especially in the treatment of grief. Part of my research team is a group of Windbridge Certified Research Mediums (WCRMs).

Julie Beischel, PhD

WCRMs are mediums who have been screened, tested, and trained using a peer-reviewed (1) eight-step certification procedure. WCRMs live in the US, volunteer to regularly participate in various aspects of research, have demonstrated the ability to report accurate and specific information about deceased individuals under several different controlled laboratory conditions, and agree to abide by specific standards of conduct including not providing readings or conveying information about the deceased to anyone unless specifically asked to. As of July 2014, there are currently 19 WCRMs who were certified as part of a research grant (there were originally 20 and one retired). All of the WCRMs and their contact information are listed at http://www.windbridge.org/mediums.htm We are no longer certifying new prospective WCRMs.

Though they are just a very small subset, I think the WCRMs probably serve as an accurate representation of non-denominational, American mental (vs. physical; see Chapter 3) mediums in general who are practicing in greater and greater numbers. Phyllis Silverman and Steven Nickman (2), editors of the anthology Continuing Bonds: New Understandings of Grief, have pointed out that when new models of grief arise in a culture, it "develops new rituals of helping to match the new model" (p. 354). They call these new rituals folk remedies and in our journal article "The possible effects on bereavement of assisted after-death communication during readings with psychic mediums: A continuing bonds perspective" (3) we propose that mediumship readings are currently serving as one of these folk remedies. The bereaved are finding that they are in need of outside assistance in coping with their grief and they are seeking out mediums as a folk remedy and finding that the information provided by mediums is well suited to fulfill that need.

Previously, I have described my research findings from 10 years of laboratory research with mediums as well as how I went from a PhD in Pharmacology and Toxicology with a minor in Microbiology and Immunology to performing research in this seemingly unrelated field in *Among Mediums: A Scientist's Quest for Answers*. In addition, I have shared suggestions for those interested in receiving a mediumship reading in the e-book *Meaningful Messages: Making the Most of your Mediumship Reading*. In those books, I was able to share my story, my advice, and my research. However, not being a medium myself, I have not been able to discuss actual mediumistic experiences or the wisdom acquired during those experiences. Their stories were not mine to tell. Thus, I enlisted the help of some of the WCRMs to share that unique information with those interested in the series *From the Mouths of Mediums: Conversations with Windbridge Certified Research Mediums*.

For the *From the Mouths of Mediums* (FMM) series, I posted each topic question to a private email listserve of which the interested WCRMs were members. The WCRMs posted their responses to the list and were able to contribute additional text in response to the others' posts. I have edited the included text only for grammar, spelling, and clarity. It was my intention to be true to the way each medium provided the information. In addition, for simplicity I have often just begun each longer direct quote with the italicized first name of the WCRM.

The collection of the posts included here should not be considered controlled research. I wanted to make sure that the collection of responses reflected a conversation among the group which is often not possible during studies in which individual participants are interviewed, surveyed, etc., and averaged or compared. In addition, I have not included here scientific assessments of the WCRMs' reports. These are their own experiences in their own words, free from the assumptions and judgments of the current scientific paradigm.

However, we have performed and published several studies examining the experiences of WCRMs (called phenomenological research). In one study (4), we compared WCRMs' experiences during a research reading for a deceased person to their experiences during a control condition in which no communication took place and found alterations in their perceptions, mental attention, and state of awareness as well as their experiences of bodily boundaries. We also noted that their experiences of time, the control they had over the content of their thoughts, and their ability to hold memories about the experience were all altered during the reading condition. Overall, these findings demonstrated that a subjective sense of an altered state of consciousness occurred during the mediumship readings compared to the control condition.

In another phenomenological study (5), we compared WCRMs' descriptions of their experiences during mediumistic readings for the deceased to their descriptions of their experiences during psychic readings for the living. Their descriptions of mediumship readings included verificatory 'signs' that contact has been established; partial 'merging' with the deceased (adopting their emotions, personality traits, etc.); apparent independence of the deceased as a separate entity from the medium; multiple sensory modalities functioning simultaneously including experiencing visual, auditory, tactile, and olfactory information pertaining to the discarnate; and the experience of 'just knowing' information about the deceased. Similarly, the psychic reading descriptions included simultaneously occurring multiple sensory modalities including visual, auditory, and tactile information about the living client; empathy with the client; and the experience of 'just knowing.' The main differences between the two types of experiences were that the psychic experiences tended to relate to the individual client whereas the mediumship experiences pertained to the deceased and that the psychic experience included information about future events (precognition) that the mediumship experience did not. In this study, one WCRM noted, "A psychic reading is like reading a book... a mediumship reading is like seeing a play."

In addition, we performed a study (6) in which six WCRMs performed mediumship readings and other tasks while their electrocortical (brain) activity was monitored using electroencephalography (EEG). As part of the study, we asked the WCRMs to intentionally enter into four subjective mental states: Recollection (thinking about a living person they knew), Perception (listening while I described details about a person they didn't know), Fabrication (thinking about a person they made up and imagined), and Communication (interacting mentally with a deceased person they knew, usually a family member). Our findings suggest that the experience of communicating with the deceased may be a distinct mental state that is not consistent with brain activity during ordinary thinking or imagination.

For more information about this phenomenological and psychophysiological research and other topics, please visit the Windbridge Institute website, specifically the mediumship research page for description of specific studies and the publications page where the PDFs of published journal articles can be downloaded.

It's important to include a bit about terminology here. In research, we refer to the living person for whom the medium provides a reading as the sitter. During our research readings which take place over the phone, the sitter is not on the call to ensure experimental blinding and scores a blinded transcript of the recorded reading (and a decoy) at a later time; in those cases, I act as what is called a proxy sitter in place of the absent sitter. In non-research settings, the mediums may refer to sitters as clients. Similarly, we refer to the deceased communicators as discarnates during research whereas the mediums may call them souls or spirits (or just spirit in general: "When I talk to spirit..."). These pairs of terms are used interchangeably here.

An additional note on 'medium-speak': I think because there often aren't English words that can accurately describe certain aspects of mediums' experiences, they commandeer words that already exist (often from physics) to describe similar concepts: for example, light, energy, vibration, and frequency. It's best that as you read their descriptions, you take these words at face value without trying to impose specific established physics definitions on them. They are simply the best existing descriptors which the mediums can use.

People often ask me for recommendations for the 'best' WCRM. However, all of the WCRMs passed the same tests, completed the same training, and agree to the same code of ethics. In our research, we're not interested in the abilities of one medium; their combined data are much more powerful. In addition, a reading is about a connection between three people (the medium, the sitter, and the discarnate) and since not all people get along with each other, a medium who is perfect for one person might not be a good fit for another. Thus, we don't recommend specific mediums and suggest that those interested use their intuition (and inspiration from discarnates) to help them decide. See *Meaningful Messages* for more information about choosing a medium.

References

1. Beischel, J. (2007). Contemporary methods used in laboratory-based mediumship research. *Journal of Parapsychology*, 71, 37–68.

2. Silverman, P. R., & Nickman, S. L. (1996). Concluding thoughts. In Klass, D., Silverman, P. R., & Nickman, S. L. (Eds.), *Continuing bonds: New understandings of grief* (pp. 349–355). Washington, DC: Taylor & Francis.

3. Beischel, J., Mosher, C. & Boccuzzi, M. (2014-2015). The possible effects on bereavement of assisted after-death communication during

readings with psychic mediums: A continuing bonds perspective. *Omega: Journal of Death and Dying, 70*(2), 169-194. doi: 10.2190/OM.70.2.b

4. Rock, A. J., & Beischel, J. (2008). Quantitative analysis of mediums' conscious experiences during a discarnate reading versus a control task: A pilot study. *Australian Journal of Parapsychology, 8*, 157-179.

5. Rock, A. J., Beischel, J., & Cott, C. C. (2009). Psi vs. survival: A qualitative investigation of mediums' phenomenology comparing psychic readings and ostensible communication with the deceased. *Transpersonal Psychology Review, 13*, 76-89.

6. Delorme, A., Beischel. J., Michel, L., Boccuzzi, M., Radin, D., & Mills, P. J. (2013). Electrocortical activity associated with subjective communication with the deceased. *Frontiers in Psychology, 4*: 834. doi: 10.3389/fpsyg.2013.00834

Julie Beischel, PhD

Editor's Introduction to Volume 1

The 13 mediums who participated in FMM Vol. 1: Experiencing Communication were: Ankhasha Amenti, Traci Bray, Dave Campbell, Carrie D. Cox, Joanne Gerber, Daria Justyn, Nancy Marlowe, Sarah McLean (which is a pseudonym for our anonymous medium), Tracy Lee (T.L.) Nash, Troy Parkinson, Ginger Quinlan, Eliza Rey, and Kim Russo. Throughout this text, I refer to each by first name. Again, if you are interested, instructions on how to get the most out of a reading are available in *Meaningful Messages* and contact information for all current WCRMs can be found at http://www.windbridge.org/mediums though I will list the specific contact information and short bios for the mediums who participated in this volume at the end, once you've gotten to know each of them a little better through their writing.

In Volume 1 of FMM, the WCRMs responded to the three questions below. As a group, they responded to one question at a time; that is, I didn't post Question 2, for example, until the conversation regarding Question 1 was closed.

> 1. How do you experience communication from the deceased?
>
> 2. What advice, suggestions, or instructions can you give to people interested in experiencing communication with their deceased loved ones on their own?
>
> 3. Why might it be that someone has not heard from their loved one when they want to?

I didn't ask the mediums to write about specific topics within each question. I just posted the question and collected the responses. In the following chapters, I have grouped similar content together. Thus, I didn't, for example, ask them to discuss their preparations for readings. I just noted that several of them volunteered that information as part of their answer, so I grouped all those segments of the responses together. That doesn't mean that the other mediums don't prepare for readings; it just means they didn't include descriptions of that aspect of their experiences in their posts. Not every topic was covered by every medium, but it seemed more coherent and meaningful to organize their responses by topic. In addition, there was some overlap across content and I did my best to categorize a given response in the appropriate section. The order of the WCRMs' responses listed here is for the most part simply logical based on the content if not entirely random.

Julie Beischel, PhD

Part 1:

Experiences of Communication

Question:

How do you experience communication from the deceased? That is, other people can't hear or see the dead when they're communicating with you, so how are you receiving the messages?

The mediums' responses to this question included three main categories: information about the structure and steps of a scheduled mediumship reading (Chapter 1: The Reading Ritual); the content as well as how it is experienced (Chapter 2: Receiving Communication); and some unique experiences that occur during particular situations (Chapter 3: Specific Cases).

Julie Beischel, PhD

❧ 1 ☙

The Reading Ritual

Like most events, a scheduled reading with a medium has a beginning, a middle, and an end. The beginning involves the medium preparing for the reading which usually includes meditation and/or prayer and/or the setting of intentions, etc., and then some housekeeping to prepare the sitter. The middle is the actual reading which begins with connecting to the discarnates and then involves the communication which is discussed in the next chapter. The end signals the closing of the experience for everyone involved.

Preparation and Meditation

Nancy noted that during her preparation, "I always ask the client to invite the deceased to the reading."

And here is what Dave, Joanne, Kim, and Traci shared about how they prepare for a reading:

Dave: "I usually will start a meditation before doing a medium reading. I ask for the spirits to come in that are for the sitter. I don't ask for information before the sitter arrives; I like to be totally fresh and not in expectation of something."

Joanne: "It all starts with intention, prayer, and protection, which is my way of 'opening the doors' to spirit. The spirit world---including my guides---now know that I am ready to work, or as I like to say, 'punching in.' I then will do a visualization and meditation and talk to my guides and ask the spirit loved ones to come and talk to me, those that would like to connect to their loved ones."

[Editor's note: Spirit guides are non-physical entities who help the medium manage the communication process from the other side. A medium may experience a team of regular spirit guides with whom she works. I think of them as spirit bouncers as one of their tasks seems to be keeping out the ruffians and ne'er-do-wells who may try to communicate.]

Kim: "First, I do a meditation prayer before I begin and I protect myself with the white light of God and I ask that any communication that may come through be for the highest good of all concerned. I ask that I be the messenger and be used as a vessel for whatever healing is necessary. Second, I ask all of my angels, guides, and spirit messengers to assist me in whatever way is necessary. If two or more people show up for a reading session, I raise my vibration to muster up more energy to help accommodate all of the deceased energies that may show up for the session. Most times I raise my vibration higher by bringing in more light."

Traci: "I ask that the reading content come in goodness for both me and for the client (and obviously for the discarnate), because I'm not about to be messing around with anything but goodness!"

Housekeeping

Ankhasha noted that prior to connecting she is "usually very alert and focused on the client, explaining how I work, and how it may be different than what they are expecting."

Nancy: "When the client arrives, I explain how I receive information so they will understand the communication as it evolves."

Traci: "At the beginning of each in-person session I turn on a recorder. I do not record phone readings but invite the sitter to record should they wish. After stating the date, my name, and the client's name, as well as my location and phone number, I then say, 'I am a medium which means I have the ability to connect with those passed-away. I am also psychic which means I am able to look at the past, present, and future. During your session I can try to do both things or focus as you wish.

What would you like to do today?' If mediumship is what they seek, it is what I always begin with. I invite questions about me, what I do, and how I do it. When that has been bypassed or exhausted, I use no ceremony or process other than to sit quietly. Information begins to flow immediately at this point."

T.L.: "I explain to each client and/or group I will be working with what may and may not occur in a reading. I, of course, encourage them to be open-minded about their session, but that a little healthy skepticism is fine. I also suggest to my client if they are comfortable, to reach out to their loved ones prior to our session and let them know they would like to 'connect' with them. However, I also share with the client that even when asking for a loved one to visit a medium session, sometimes a 'connection' doesn't occur. I further go on to explain that this doesn't mean their loved ones aren't interested in speaking with them or offering messages, but that our connection---spirit's and mine---is simply not strong enough. I will ask my client if they would like to reschedule and attempt to try and connect at a later date and time. If they agree, we hold another session and if I still can't connect I suggest they may want to seek out another medium, as every medium is different and they may have a better experience with another reader."

Connection

Regarding making the initial connection, Dave stated, "When the sitter arrives I sit across from them and start to open up with the white light."

Carrie described her process as such: "I perform readings by tapping into what I call the 'Love Energy.' I have to find some 'love' connection, either for the person (sitter or discarnate) or between the individuals. This makes reading for people who had tension between them difficult or for individuals who have no emotional connection more difficult. This makes participating in the controlled environment of Windbridge Institute research readings unique for me because the researcher has no 'love' for the discarnate, so the energetic experience is slightly different for me. I try to circumvent the researcher and tap into the love between sitter and discarnate, but it makes the energy feel further away."

Joanne noted, "I find that for a family sitting together for a session will tend to strengthen the energy of connection. I have also witnessed this during many of my spirit communication demonstrations. A particular group of family members that sits together would tend to receive more messages from their loved ones as opposed to those who had attended the event individually."

Ending

Several of the mediums specifically described how a reading ends in their practice. Dave stated, "When the images and impressions stop coming in, the spirit is usually done. I will ask the sitter if they have any questions before totally ending the session." Joanne noted, "Afterward, I always thank my guides, angels, and loved ones, officially 'punching out' for the session/day." Ginger wrote, "After the session is over I feel a sense of deep love and sometimes ecstatic joy as the energies of the discarnate leave me and I become myself again."

❧ 2 ❧

Receiving Communication

This chapter focuses on the content of the information the mediums experience (the 'What') and the manner in which they experience it (the 'How'). Often, the type or content of the information will determine how it is experienced so there is much overlap between the What and the How. For example, in experiencing a discarnate's name, specific memory, favorite food, and cause of death, the medium may mentally hear, see, taste, and feel, respectively, that information. I have attempted to separate the descriptions of the What and the How below for clarity, but that separation was often tricky, as you will see, so we will begin with some examples of their intersections.

Ginger: "I am shown images of things the discarnate did in life, I hear the music they liked, taste the food they ate, see where they spent a lot of their time (i.e., work, play, home, with family), I am shown where they are now in regard to the sitter, and I feel what they felt before they passed. They always want me to experience their moments before dying so I feel the pain and what was happening before they passed as well as their awareness of what transpired before death."

Joanne: "I get a sense in my body of how they passed, or I may see within my third eye some sort of movie or still picture and objects that will show me information connected to their passing. I will hear names and dates, see numbers and letters and places, feel their personality, get a sense about their life with their hobbies or career, see memories past connected to the discarnate or the sitter, see objects that are connected to the deceased or have meaning to the sitter(s). Sometimes I get a sense of smells or odors, or am able to 'taste' foods with special connections with the discarnate/sitter (most likely because I am a foodie). I remain in a heightened state of awareness, continually observing my senses, while relaying literal and symbolic communication to the sitter(s) until the session is over."

Nancy: "I can be seeing the discarnate and at the same time I can hear the discarnate speak to me. In the meantime, I may smell their favorite perfume or sense their depression or feel the blow to the head that killed them. During a reading these abilities can occur separately to bring in information, or they can occur all at the same time."

The What:
The Content of Mediumistic Experiences

In *Meaningful Messages: Making the Most of Your Mediumship Reading*, I describe the three major types of information reported most often by mediums during readings: information that helps the sitter identify the discarnate ("It's me"); information that describes events in the sitter's life that have occurred since the death ("I'm here"); and messages specifically for the sitter ("I love you"). These and other types of content included in mediumistic experiences are described below.

Identifying Discarnate Descriptions

Joanne: "I can then get a sense of about what age they were when they passed, their hair color or hairstyle, clothing/fashion, and try to see the color of their eyes. Sometimes this information comes sporadically throughout the reading. For example, I may see the beautiful blue eyes at the end of the session but can describe the age and physical description at the beginning of the session. Sometimes when I receive communication from a quieter energy, I may need to close my eyes to really focus and talk to them to ask them to talk to me. Others come forward full force if that's the way they were when they were here on the physical. From my experience, personality does not change from the physical to the spiritual."

Dave: "I frequently will get their passing to identify who is coming through and the personality of the deceased. Once the connection is made and the spirit is identified the information seems to come so much quicker and clearer."

Kim: "I will always ask the deceased energy to give me details to prove it's them. At that point, they usually will spell a name (their own name), the person's name whom they are with on the other side, and sometimes if they want to talk about a living relative or friend, they will spell that name above the sitter's head. First, middle, and even last names can come depending upon how good of a communicator the deceased soul is. When they make numbers appear above the sitter's head, it often indicates the person's birth date or the person's death date. I ask many questions of the deceased to prove their identity. It becomes like an interview. Sometimes they ignore my questions and have their own agenda. In the end, my guides will always make the deceased energies prove their identity to me however they feel the sitter will be impacted the most."

Nancy: "I begin each reading asking spirit to show themselves to me. I begin to start seeing the spirit. They usually begin by showing me their most identifying features. If they had beautiful teeth and a great smile, they show me their smile. If they had gorgeous, expressive eyes, they look at me and I see only their outstanding eyes. As the revealing of physical features continues on and on, each body part is revealed, one right after the other. Sometimes, I see the whole body from the top of the head to the shoes they wear. Sometimes, I only see a face or maybe their hands. When I see the whole body, I see through them. They are transparent, like smoke or sheer fabric. They show me how they died and they show me what it was like to cross over."

Ankhasha: "Sometimes the person's face appears first, so that is what I describe, sometimes it is a special piece of clothing they wore, a uniform, etc. Sometimes I can see a clear image of the spirit, and I ask them to show me how they died; usually that opens up the feeling I get in my body as to how they passed. Just recently I was starting a reading and when I asked that question my head snapped suddenly to the left. The deceased had hung himself. A few years ago I asked that question and felt like I was in one of those 'quick drop' thrill rides they have at amusement parks. The person had died in an airplane crash; that was an unusual feeling. And sometimes I get nothing at all."

Messages

Dave: "Messages are often given to the sitter, sometimes regarding future events or what is currently going on with the sitter."

Ankhasha: "Feelings of regret, anger, guilt, confusion, happiness, and love all come through; at least it has been my experience that they aren't all messages of love. They seem to be as varied as people are."

Joanne: "Once the spirit loved one has provided sufficient evidence of their identity, they will often let the sitter know that they are okay and may acknowledge who helped them make their transition to the other side or wish their grandchild a happy birthday or express their knowledge about an upcoming event in the family such as a graduation or marriage."

Kim: "I will get many emotions running through my body of the deceased energy; i.e., anger, resentment, sadness, regrets, and especially love. The emotion of love comes to me in the strongest way."

Ginger: "I am told how to move or talk as the interaction takes place. I use the discarnate's inflections and am directed to move my body and hands a certain way when speaking for the dead which resonates to what the sitter is experiencing in that moment with me. It is powerful and makes me feel different for a bit as the communication often turns to the sitter receiving very loving and often times emotional messages from the spirit interacting with me."

Symbols

The WCRMs often described receiving symbolic information requiring interpretation and the process of learning that vocabulary of symbols. In research, we instruct them to "say what you see" to convey the symbol and then provide an interpretation for it. Often, the experienced symbol can be meaningful to the sitter but not the medium's logical interpretation of it, so we try to collect both.

T.L.: "If I see symbols they will vary and their messages will differ. For example, a white bouquet of flowers from spirit means 'I'm sorry' but a pink bouquet of flowers will represent 'happy birthday' to acknowledge a birthday which just passed or is coming up. This also can be followed by a number to signify the date or month of the birthday. An anchor on the upper arm of a person will represent a military connection, while an arm shown from the elbow to the hand and holding a drink will represent alcoholism or some other addictive propensity."

Sarah: "The information except for emotional and physical feelings, smells, and hearing comes in visually as symbols. And often I need to translate the sentiment of the symbol as it bears meaning for the deceased and how it might have meaning for the sitter. I assume that the deceased is using my database of past experience and in a sense trusting that my dictionary and their symbol can be made understandable to the sitter. If not we keep trying. It's interesting that when translating a symbol, it feels like unspooling thread; the symbol is heavy with meaning and as I speak the meaning and its sentiment unravels until there is little tension, most of the time by spool's end, the sitter understands what the deceased is referring to. Sometimes the symbol is weighty in its simplicity because it is meant as a sign from the deceased in confirmation to the sitter. In these instances simply passing on what I see is the important part, not necessarily making meaning from it."

Traci: "Sometimes names will come through with a symbol, such as Mickey Mouse for Mickey/Michael/ Mike/Micah/Michelle/ Shelly. When a floral name is coming through, I might see it specifically as that flower such as Iris, Lily, Daisy, Delilah (which comes through as a dahlia), or Violet. Sometimes with violets if the name does not resonate, it otherwise will indicate the person favored or raised African violets. When Rose (Rosa, Roseanne, Rosalie, etc.) comes through I may see a rose and/ or smell roses. If a certain flower was actually draped in a blanket over a casket or grave, I will see that flower in a blanket formation. When lilies come through, they also are shown as a fleur-de-lis emblem. Quite frequently seeing the fleur-de-lis is indicative of a profusion of lilies at the funeral. Too, I have a kitty named 'Lily' and in my mind's eye, I will 'see' my Lily, which suggests the name Lily is important. As strange as it may sound,

Lily-the-cat meows at opportune times during readings, generally indicating a message needs to be reinforced or expounded upon. This is a sense I've learned to follow, and experience demonstrates Lily to be consistent! Concerning names, not as often deceased people from my personal life will show themselves, in a flash. My father was Dwayne Merlin, both uncommon names; he also was an ironworker, brakeman on a railroad, volunteer firefighter, and a barber. When he appears I pull from these characteristics as well as his names. Finally on flowers, my maternal grandfather George was a prolific gardener. When he pops into a reading it is indicative of a person named George, the maternal grandfather figure, or of a dedicated gardener. As compared to when I first began reading for others, the 'language' if you will, of the deceased is much broader today, ever-developing. It truly has been like learning a new language. As an early reader, spirit helped me to build confidence I believe by occasionally showing a hammer (like what you would pound a nail into the wall with) dangling in mid-air. It would tap a few times. I learned this meant that if a client was resisting the message, to repeat it to them. When I repeated it, at times the client would then acknowledge the issue at hand. If they did not, the hammer would again appear, yet pound harder and faster! As I learned to trust the hammer, eventually it rarely appeared. As I reflected on its disappearance, I realized that I had learned to automatically trust what I was seeing, as well as to trust my sense to again approach an item that the client rejected initially. The hammer was a fun thing for me to have the opportunity to work with, to learn from. My environment while reading also offers insight into reading content. Upon occasion my television might be on while I do a telephone reading, yet muted. While doing the reading I likely am oblivious to the television, yet might inexplicably be drawn to the picture. If the picture is of a house fire, perhaps the discarnate experienced a house fire, and if it is a segment on combat in the Middle East that I'm drawn to, the discarnate I'm in connection with likely has a connection with the same: a living child that served in the military, stationed in the Middle East, for example. I have learned to trust that the weather, media, outdoor noise (such as sirens), dogs barking, often bring messages that slip right into the reading accurately. Thus, paying attention to my environment proves to offer relevant information."

The How: The Manner of Mediumistic Experiences

As described in the introduction, during research we have been able to capture the multi-modal nature of the way the WCRMs describe mediumistic communication. That is, the information is experienced in the mind through the mental 'senses' including visions (through what in parapsychology is called clairvoyance, basically translated as 'clear seeing'), sounds (clairaudience), smells (clairolfactance or clairalience), and tastes (clairgustance). The experiences may also include 'just knowing' (claircognizance) and feeling the discarnate's emotions as well as pain or other physical sensations related to the discarnate's cause of death or other health issues (clairsentience). Below are descriptions of the WCRMs' mental 'sensory' experiences.

Sensory Experiences

Ginger: "I smell, see, taste, hear, and feel them as though they are a part of me."

Nancy: "The visions appear much like a movie with one frame appearing one after another. Sometimes the movie runs fast, and sometimes the movie runs slow, much like an old slide projector."

Eliza: "During the reading, the information from the deceased comes through either with visions/pictures or words. It depends on the deceased's chosen method of communication."

Joanne: "I hear, see, feel, smell, and taste things as I am receiving the information. I see colors in my third eye. I will then focus on listening, watching, and feeling the energy of the discarnate. Sometimes I see them with my eyes open, and other times I see them with my eyes closed."

Carrie: "I typically can see a physical image of the person in my mind's eye like dreaming with my eyes open. I can see the discarnate in the room but also see the backdrop of the room around them. Then I start to feel the information through my body. The feeling translates into 'words' or a communication of ideas. I will also have physical sensations of smell and taste, sometimes pain, and in one instance I experienced going deaf, in another I threw up."

Dave: "I receive messages in multiple ways sometimes in one sitting; in any given session, all of the senses may come through. One is through clairvoyance; I mostly see what I am getting, the pictures they show me in my head: an image of a person, scenes, etc. I hear words, feel the energy and emotional state of the discarnate, and sense the personality of the individual. Once in a great while I will actually see visibly a discarnate in the room. One time I saw a lady's husband behind her for the entire reading! I could even see details in his face clear enough for me to describe to her."

Troy: "Spirit communication comes forward through me through clairaudience, clairsentience, clairvoyance, or a combination of all three. I see them within my mind's eye. I often tell people that I see spirit the same way as if I told you to imagine Marilyn Monroe standing by you. You wouldn't see her physically like the table or chair but you would see her within your mind's eye. I hear spirit telepathically using clairaudience, meaning within my mind. I don't hear a whispering or an actual voice outside of me; it is more a conversation that occurs within my head. However, with clairsentience, I actually do physically feel pain or a temperature change within my body. It tends to occur as a twitch or a heat, which indicates the spirit may have had an injury or an ailment in this area of their body."

T.L.: "Typically I will 'see' images in my mind's eye, such as a movie playing out on a movie screen or I will see things literally written down in my mind's eye such as words, numbers, and other symbols. I will 'feel'/'sense' things as well. If I am feeling/sensing something it can be in the form of goose bumps, an electric feeling or a buzzing feeling which will cover my entire body or sometimes just at the top of my head, or a warm sensation that permeates all over. If spirit is speaking of something emotional that was a challenging experience endured by spirit when they were living or maybe by a living loved one who is going through something difficult, my chest hurts or has a very heavy feeling to it. If spirit had a great sense of humor I will feel 'light' or maybe see The Three Stooges or something else funny that will indicate to me I am dealing with a 'comedian' of sorts. Smell can also play a role, so if they smoked or drank, for example, I will smell it."

Ankhasha: "Sometimes I see things in a movie format, an entire scene runs in front of me, other times I see only a flash, like a subliminal advertisement: They come through visually quickly and clearly; like a flash, but very clear, over in an instant. When that happens, it is very choppy, hard to get a hold of the entire picture. Sometimes I see them in kind of a fast blur, hear them loudly, but don't really feel any emotion from them unless I spend time with them. It has been my experience that the ones who are able to stay around for longer times during the reading make their presence known by almost a building of energy, as if they are coming closer and closer as they communicate with me, until I can hardly hear anything, the sound is so high-pitched and loud and there is a buzzing, humming glow that becomes hard to look at. It almost feels like I am being lifted, levitating while I am communicating with them. I know that may sound wacky, but that is what happens to me. And to be honest, it feels really good!"

Traci: "Information comes to me via the gamut of senses: hearing (it may be a name, a particular 'saying' or accent, an animal, a cry, a speech idiosyncrasy, the wind, a crash); seeing (can be a symbol, a still as in a photograph, or a moving scene like watching a vehicle accident occur; also communication comes with words via a marquee, or in reading a page placed in front of my mind's eye; the typeset can be significant, or the design of a letter: Victorian versus a technical-type of font can be indicative of a number of things); smelling (may indicate anything from a favorite or detested food; a perfume; or, if a flower such as a rose, either the name Rose/Roseanne/Rosalee, etc., or the discarnate loved or grew roses, for example); touching/feeling/being touched (too at times I experience shivering on top my head or down my neck or shoulders or back; this is an indication to me that the discarnate is letting me know I am on target); tasting; and 'just a sense.' It is important that I pay attention to first-thought as in: what comes to me powerfully, initially, and to not bypass it. Generally in readings, all of the above mentioned 'senses' come into play within each session. I also experience sympathetic pain particularly in regard to cause of death. Examples of this include an explosion of pain in my head indicates a gunshot to the head, whereas a sudden slap of movement with pain to the head may indicate a vehicle accident with head injury. In contrast, a sudden dart of pain may indicate an aneurism, or a throbbing pain or localized pain in head may indicate migraine, cancer, or tumor."

Verificatory Signs

The phenomenological research described in the Introduction found that another common experience during mediumship readings was a 'sign' verifying that the medium is in contact with the discarnate. The WCRMs who participated in FMM Vol. 1 echoed this experience and discussed similar specific sensory signs.

Dave: "I feel a tingling at the top of my head and around my shoulders when I/they are ready."

Eliza: "When a message is emphasized, I often feel a sensory shiver (that's the best way to describe it) up my spine. Other times when the deceased wants to reinforce a message, I see it in 'lights' around the word or words; sort of like a neon street sign with lights around."

Traci: "When a discarnate is coming through with a name, at times an arc appears below or above them. If above, it means that the discarnate who is with me shares a same name (first/middle) with someone older than them. When the arc points below them it indicates a child, usually their own child has the discarnate's name (first/middle). Arcs in both directions indicate first/middle names being passed through families or a child carries the mother's maiden name as a middle name. The arc is a white line, as if written in pen, scrolled in my mind's eye."

Placement in the Sensory Field

Several of the WCRMs described experiencing different types of information in different 'spaces' in their visual field or physical space.

Eliza: "I see the deceased around the sitter and often the mother's side of the family presents on the sitter's left side and the father's side of the family presents on the right side. If someone had a heart relationship with the sitter, such as a husband or lover, I see them behind the person. If it is a friend or cousin of the sitter, I often see them standing a few feet back from the sitter."

Kim: "I will see a person in my mind's eye, male or female, and they show me their relationship to the sitter (usually based on age). So if it's a mom or mother figure, they will show me one generation up in age from the sitter. If it's a grandma, they will show as two generations up in age, etc."

Joanne: "[At the beginning of a reading] I will meditate until I feel energy shifting, which usually feels like it is coming in on my left side and have the sensation that I have to turn my head to the right, to hear them better."

Timing

The speed and timing of experienced information was also discussed by several WCRMs.

Eliza: "Quite often prior to a scheduled reading, I have a visitation from the deceased of the person (sitter) I am reading for. The visitation will come in a dream, a feeling as if someone is standing next to me, or a vision in meditation."

Nancy: "Sometimes, they begin making their presence known before a reading. I start sensing them draw near me. I begin feeling their energy, mood, excitement, or desperation to communicate. Often, they begin to speak a word or two here."

Ankhasha: "When I am doing a mediumship reading for a client I prepare for them by asking any spirit who wants to contact them to make themselves known to me. Many times they will show up in my dreams or just appear the week or day before the reading. I look at this as kind of a 'trial run' on their part, and I will start to receive information for that client prior to the reading. That information I usually write down as I may forget it. I always feel that during those types of visits I am in a daydream."

Joanne: "The information comes very quickly, so I like to have a pen in my hand and start to write once I feel the energy of the discarnate starting to develop."

Traci: "Communication is immediate for me when sitting with a client, from the discarnate."

Dave: "Names, dates, ideas, pictures or scenes, sometimes smells, may run through my mind, and I try to verbalize them as quickly as they come. Sometimes a word will just pop in my head and I say it."

Other Descriptions

Lastly, below are some additional general descriptions about mediumistic communication as well as topics that didn't fit nicely into the What or How categories but that I thought were important to include.

Ginger: "When I receive communication, it is a lot like having a good friend come to my house, knock on the door to get my attention, and then walk in. I see them like they are in the room with me and they talk to me like a living person would. When communication begins, it starts with that sense of someone knocking on my energetic door to get my attention. It is not scary or painful; it is rather soothing. I feel a sense of joy at the prospect of getting ready to communicate because the energy that I feel from the dead is so clear and usually very loving. As the communication begins, it is a very intense awareness of the person who is standing next to me, like someone is in my personal space bubble."

Daria: "The best analogy I can use, in describing how I receive messages from the deceased, is to imagine you are having a daydream. One day, while I was sitting at my kitchen table, I found myself daydreaming about a man in a gray uniform with a delivery truck filled with ice cream and snacks. In that instant, I was able to see an entire scenario of images in my mind's eye. Later, while speaking with a close friend, I relayed the daydream because it was such a unique experience. She listened and then burst into tears; the man was her father. He was an ice cream delivery man who wore just such a uniform. That was the first time I realized daydreams were vehicles for messages."

Carrie: "The way in which I receive information is as much about the skill set of the discarnate as it is about me. Some discarnates come through like gang busters and their energy comes strong and fast. Others have energy that ebbs and flows in waves like tuning a shortwave radio; you can hear them trying to get the right frequency and it is harder to understand. To put it another way, I am like an instrument that they can play based on their skill set. One discarnate may be able to play like Beethoven and another may only be able to play "Chopsticks." I recognize that we have skills that allow us to open to their communication, but they are learning, too."

Sarah: "Contacting the deceased is like finding a specific frequency on a radio: the word vibration is pretty accurate because you are looking for a pulse or presence that is not your own. I use both my mind or psychic sense and what feels like a beam of light from my chest that acts like a flashlight. This allows me to scan and feel in order to feed back information to my mind or psychic sense for translation."

Kim: "Once the reading begins, I take a deep breath and wait for the movie to start. The energy of the sitter projects movie-like images that are conjured up by their deceased relatives. I interpret what I'm seeing based on my own frame of reference. My spirit messengers will show me the movie based on things I will recognize and understand from my own life."

Ankhasha: "This is the way names come through for me: either I am told the name right away as soon as the spirit makes themselves known to me, or I use a technique my beloved friend psychic medium Annette Martin taught me. She told me years ago to keep a notepad in front of me during the reading and not concentrate on what I write, just let it flow: doodles, initials, words, etc. Since she always used a yellow legal pad, I figured it must have magical properties, so I started using one, too! (OK, you know I am kidding, of course.) I enjoyed the process so much and found that before long I was writing down names, places, important numbers, events, etc., and studied more about automatic writing. At times the handwriting will change depending on what spirit is coming through. So now I always try to have a yellow legal pad in front of me during my readings."

At a different time in the online discussions, Ankhasha wrote, "Over the years I have learned to 'trust the message' even when it makes no sense at the time to either myself or the client. I don't like filtering information or trying to make sense of it, because every time I do that, I can end up in the wrong place. I used to receive a lot of pictures, symbols, visually; however now I seem to repeat what I hear from them. But I can't really say I am repeating it since most of the time, the first time I have heard it is when it comes out of my mouth."

❦ 3 ❧

Specific Cases

Several of the WCRMs also noted that some experiences occur only during specific circumstances related to: whether the discarnates are children or animals; the beliefs of the discarnates and whether they perpetrated emotional or other damage during their physical lives; and the level of the medium's trance. The mediums also discussed the necessity for setting boundaries with discarnates.

Children and Animals

During research readings for specific discarnates, the WCRMs often spontaneously describe associated deceased companion animals (pets; for example, "she has the little white dog with her") and even infants and lost pregnancies (for example, "he is holding the baby girl"). At the time of this writing, we have begun collecting data for a study (1) examining the accuracy of readings for specific deceased companion animals by WCRMs.

Kim specifically discussed how she experiences children and animal communicators differently than adult humans.

Kim: "I find when children come through they tend to show me a movie in color with playgrounds, animals, and all of their friends and adult protectors. It's a much more innocent vibe and the movie looks like it is in 3D with vivid colors. Animals come to me as well but are usually accompanied by a human soul (usually a relative of the sitter)."

Beliefs of the Discarnate

Carrie: "Some discarnates appear to be saddled with the same beliefs they had during life. I have had discarnates who were very religious in life decline wanting to talk to me because of their belief systems in the afterlife. I don't know if they truly still believe these things after death or if it is just an identifying factor for them, a way for their loved ones to recognize that it is really them. I have also had the experience of the discarnate who will come through begrudgingly and say, 'I didn't believe in psychic mediums in life and I am not really keen about doing it now.' They will usually still give me information for their loved one, but I can tell they are holding back or uncomfortable with the experience."

Ankhasha: "I tend to connect with the deceased regarding their relationship with the sitter emotionally, how they were feeling upon their death, and, if it has changed, how they feel now. It has been my experience that not everyone changes or has more awareness over there."

T.L.: "Often times, if spirit has changed their beliefs and how they lived in life differs from how they are in spirit form, I may be shown a split screen with differing emotions, feelings, sensations which are heavy, difficult, or challenging on one side of the screen followed by the other side of the screen whereby spirit might be in a setting that is more ethereal, a soft glow around their energy, etc., that will indicate they have changed their beliefs or raised their vibration."

Hurt Inflicted by the Discarnate

Eliza: "Many times when I am doing a reading for a person whose loved one abused them while they were here on earth or if the deceased is not sure if they will come through because of an action they did, the deceased person is usually not standing close to the sitter but either about 4-5 feet behind them or to the side behind this dark curtain peeking their face out. I tell the sitter who I am seeing and ask them if they want to connect. Quite often the loved one will come forward and communicate if the person expresses some form of resolution and/or forgiveness."

Daria: "I have connected with deceased souls who behaved so badly in life that they are ashamed to show themselves when the opportunity arises. Once I help everyone involved understand that from a higher perspective there are no victims or villains, just souls who incarnate together to learn and grow, that shame is transmuted to a willingness to forgive that brings soul growth and an openness to communicate."

T.L.: "If they struggle with something in their spirit form that they also struggled with when they were living (for example, they were unforgiving or rigid in thought), I will feel frustrated or may see hands clenching something tightly indicating that letting go or forgiveness is difficult. I may see a narrow tunnel with little light representing narrow-mindedness or rigid thinking. If they make themselves known by their presence in a reading but don't feel like speaking, they tend to stand far back, kind of like being in the shadows. They want to be recognized, perhaps, but not enough to speak at that time. If this happens it is usually because they are uncomfortable with how they behaved in life and are still trying to come to terms with the damage they did."

Trance Experiences

In general, there are two main types of mediumship: physical mediumship in which physical effects such as table-tipping, direct voice, materializations, lights, etc., are produced, and mental mediumship in which information is communicated to sitters. The depth of trance during any given reading by any given medium may vary widely from a fully conscious but somewhat altered waking state (which is what occurs during research and what you'll most likely encounter as a sitter) to a fully unconscious trance state. The WCRMs who participated in this volume of FMM experience a continuum.

Ankhasha: "There are times that I am very alert and tell the client what I am seeing, hearing, feeling, etc. But now most of the time, if I sense that they are open to the experience, I do enjoy allowing the spirit to move into me and use my body to speak with the client. I prefer trance work (the deeper the better) whenever possible for a number of reasons. It does seem to energize me, but most importantly, with me 'out of the picture' so to speak, a very clear message comes through (unlike my ramblings here!). Trance work is also my preferred way of doing a paranormal investigation; if I can contact the spirit that is there (if there is one) then I allow them to talk with the investigators. I am never afraid of allowing something to 'possess' me; I don't encourage or entertain fear. I am still in control of the process and can stop it if I want to."

Kim: "Sometimes the deceased can use a combination of showing me a movie in the reading as well as allowing me to become them. I'm not in a trance but I do not remember what I said after the reading. I won't let the deceased take over my body, but I do become them as they are trying to relay messages. I become them with their emotions intact (including their opinions), their visions (the movie), and they very often will show me a movie of what it looks like where they are. For the most part, when a spirit or deceased energy is at a beginning level, I will see them in a blank room with no walls, just vast space. I believe I see it this way so there are no other distractions and it helps keep me focused on the spirit."

Nancy: "Once the spirit feels comfortable with me and I feel comfortable with the spirit, I allow the spirit to enter my body. At that time, I see the world through the eyes of the spirit. I hear their thoughts. I behave with their personality, and I move my body the way they moved when they were alive. In other words, I become the dead person. People who see me do this are often taken aback. I've been asked if I am an actress because the personality of the spirit is so different than mine. Once the dead person is in my body, I can look at 'my' arm and see what color my skin is. I can see the clothes I am wearing. I can see what the spirit is seeing. Their personality becomes my personality. My tone of voice may even change. If they were an artist, I begin to 'sling paint with a cigarette hanging out of my mouth' just like they did. If their death is a mystery, I can see what happened to them because I am looking through their eyes. If they died on the operating table, I can see why (because I am literally in the room at the time of death.) Having a spirit enter your body is like having two separate people in one body. Who is in charge? I am. It is my body. I am very aware of my own thoughts and the spirit's thoughts as being separate. I can feel their presence within me as it presses outward against my skin. I allow them to communicate in this way as long as it is comfortable mentally and physically for me. The spirits don't usually stay in my body long once they enter. (It must take a lot of energy for them to use my body in this way.) They come in and go out usually within a 2- to 6-minute time period."

Setting Boundaries

During the 8-step screening of WCRMs, they answer an interview question asking, "Are you always receiving communication or do you have to tune in?" The vast majority said they had to tune out in order to live a normal life (and get the dishes done and walk the dog, etc.) or they would just be experiencing communication all the time. As we as scientists and we as a society gain a better understanding of the normalcy of the mediumship experience and its specifics, I predict that this will be an important skill we teach to young mediums: setting boundaries for discarnates.

Ankhasha: "When spirits appear they can just 'pop in' unannounced; I can be shopping, talking on the phone, even watching a movie or reading. When they do that I will ask them what they need, make note of it, and then thank them. If they are too persistent then I have to set boundaries."

Nancy: "In my everyday life spirits come to me uninvited. They just show up. I feel them press against me wanting attention. I hear their voices or their laughter or their music. Often, I see them standing in the distance. They stare silently at me while telepathically asking for permission to communicate. Since they know I can see them and hear them, spirits seem to always be around me. The constant activity is very tiring for me. So, I have learned how to politely put up a 'closed shop' sign on the door to communicate with me. The truth is, some spirits are very persistent and come in through my dreams or catch me when my mind is preoccupied with something else. These are usually souls who want justice or an understanding of their actions on earth."

Joanne: "We are always in control of the communication. When I was young, I didn't know how to 'turn it off.' As we develop as mediums we learn how to set boundaries. Prior to an event or session, I ask for spirit loved ones to communicate with me. Sometimes they will show up a day or even at an earlier event, but this is more rare than common. I don't take people's spirit loved ones home with me after an event. They are more interested in hanging out with their family than hanging out with the medium!"

References

1. Beischel, J. (2012, June). *Anomalous information reception by credentialed mediums regarding non-human animal discarnates.* Presented at the 31st Annual Meeting of the Society for Scientific Exploration, Boulder, Colorado.

Julie Beischel, PhD

Part 2:

Advice for Communication

Question:

What advice, suggestions, or instructions can you give to people interested in experiencing communication with their deceased loved ones on their own?

I often assert that if one human brain can experience communication with the deceased and report accurate information, then we all have that potential. Granted, by that logic we all also have the potential to paint masterpieces and play in the NBA. Unfortunately, there are limitations to the actualization of our potential, but that doesn't mean we can't practice and exercise and optimize the potential. The WCRMs agree with that assertion to varying levels.

T.L. notes, "It's not something specific only to certain people. It is something we all can do. While the degree of what each of us is capable of can vary, I do not believe that only certain people can speak to the other side. I believe it's something we all can do and this is something I share with my clients in every reading." Traci stated, "It is not my experience that everyone in our world is equipped to pick up on signs or to believe them." Nancy says, "While it is true that we are all born intuitive and many people receive communication from the dead, the truth is that many people do not. I believe this the norm, not the exception."

The WCRMs who participated in this FMM volume provided thorough advice for those interested in experiencing communication. The following chapters discuss the WCRMs' general advice (Chapter 4), the types of communication you may experience (Chapter 5), and specific suggestions for optimizing the experience (Chapter 6).

For clarity, I reworded some of the WCRMs' suggestions as instructions. So if they wrote, "I usually tell people to trust that communication is possible," it is included here as "Trust that communication is possible." I also changed some general statements (for example, "spirit can communicate with them by...") to more personal ones ("spirit can communicate with you by..."). Please do not "hear" the WCRMs as bossy know-it-all's; they were polite and gentle and, for simplicity, I reworked their words into the demands of drill sergeants.

⚜ 4 ⚜

General Advice

Many of the WCRMs reported that they are often asked by clients about how to communicate with their deceased loved ones. Carrie noted that, "I get asked this question by almost everyone I read for" and Nancy called it "a universal question. It is one of the 'big ones' everybody wants to know."

Several WCRMs discussed how one of the main goals of their practice is facilitating independent sitter-discarnate connections. I've even heard some of them convey the following sentiment: "My ultimate goal is to put myself out of business because everyone is able to communicate with their loved ones on their own." Below are some additional thoughts along these lines.

Carrie: "My goal as a psychic medium is to help them create their own personal connection with their loved one, so that they no longer need me, and can connect on their own, in their own way."

Ankhasha: "I tell my clients that although my purpose is to bring forth validating information from their loved one, they need to know that they can make the connection themselves. I tell them they can do this through allowing themselves to grieve their loss and then by opening up to new experiences and situations around them that give them signs that their loved one still exists. I do not encourage clients to visit me too soon after a loss (although I have allowed that to happen a few times) or to book another session after the first one, unless for some reason I am unable to make a connection for them. I do not want to encourage dependency on me or any medium as they work through their grief process."

Daria: "Something that gives me great satisfaction is helping people understand that they don't need me to communicate with their loved ones; they just need to ask, be open, and trust in their ability to recognize when they are being 'nudged' by those souls with whom they share a bond. Trusting the validations you receive from loved ones makes it easier to be receptive, open and learn the subtleties of working with energy."

Before we get to the types of information you may experience and specific suggestions for receiving communication in the next chapters, below is some of the WCRMs' additional general advice.

T.L.: "Clients will often times ask how their own communication with loved ones can occur and the greatest advice I can give to them is to believe that communication can actually take place and then be open/receptive to the communication occurring. Believe that you can communicate with your loved ones."

Eliza: "Believe any thoughts or intense emotions regarding memories and dreams about your loved ones are your loved ones attempting to either just hang out with you or give you a message."

Joanne: "The energy of love is the bond between the physical and spiritual worlds. Upon the change called death, love transforms, but it never dies. Your love for your loved ones passed is your connection to them. The key is to release expectations, be patient, and trust. Release the thought that it is your imagination or a coincidence. It's not!"

Nancy: "It is the dead who must make contact with the living. It is up to the living to recognize that contact when it is made. So you can try this and try that. You can study this and study that. The truth is, after you have learned all there could possibly be to know about communication with the dead, in the end, the ball still remains in their court."

Ginger: "My best advice would be to totally trust that you in fact can communicate with the deceased and then trust it when it happens. I see people try so hard to make that connection; they do meditation, they go to many mediums for advice on how to do it, they read books, and they try and try, when really all they have to do is be open to communicating, watch for the message, and then trust that it came."

5

Types of Communication

In Part 1 of this volume, we saw the many variations that the WCRMs described regarding the What and the How of their communication experiences. Nancy noted that "the methods dead people use to communicate with the living are as varied as the personalities of the deceased." Troy stated, "Be open to communication coming through signs, symbols, music, and dreams." Below are some general categories of communication and situations conducive to contact for which you may wish to at least keep an eye out if not intentionally induce.

Signs

Ankhasha: "Many times in the reading the deceased will give some kind of a message relating to a sign that they will send, and I tell them to look for it, and thank their loved one for sending it to them."

Traci: "Signs often include a phone ringing, a dream, a song on the radio. I encourage people to pay attention to animals, clouds, and weather. Too, I suggest that they be open to electronic idiosyncrasies such as flickering lights or the stereo turning on by itself. And that the fly that shows up every family dinner on the chandelier even if it's 25 degrees outside is worthy of note. If you are open to believing your loved one's true presence in a dream, in a song playing on the radio, in electronic interference, a random text message, a scent wafting in front of you, the shape of a cloud, feathers falling, etc., these are ways loved ones visit."

Daria: "Some messages are very subtle, and they can come in a variety of ways: hearing a name at the perfect time, turning the television on to a significant show, having them come to you in a dream or having dates or numbers reappear over and over. I once told a man that his mother was showing me a butterfly for him. Later that day, he was sitting on his porch, relaying the details of our reading to his family, and when he got to the part about the butterfly, one landed on his leg, exactly on cue. The butterfly continues to come as a symbol when his mom is around."

T.L.: "I encourage my clients to ask for signs from their deceased loved ones. Asking for a particular song they may have shared and hearing it on the radio the following day, seeing a butterfly with a specific color, finding a dime or a penny, feathers, even numbers such as those found on license plates can double as messages corresponding to a birthday or a wedding anniversary, etc."

Joanne: "The signs from spirit loved ones can be very subtle because they are now pure energy and are now communicating with us without the aid of the apparatus of a physical body. Spirits can use small, easy-to-move objects like jewelry and coins to communicate their presence. After the passing of a loved one, many will frequently find coins of a particular denomination such as pennies ('pennies from heaven'). Signs may come at moments when you least expect them and in the strangest of ways. If you remain open and aware, you will become more aware of signs."

Dreams

Joanne: "Before you go to sleep at night, you can ask a spirit loved one to come and visit you in a dream."

Ankhasha: "Ask for your loved one to visit you in your dreams; this seems to be easier for some to access that realm."

Daria: "Sometimes people say they only communicate in dreams and don't believe it's real. Understanding that our loved ones will take this opportunity to connect with us and help us resolve issues can only help if we trust the process."

Eliza: "Keep a dream journal, since quite often passed loved ones visit us in dreams but people forget when they wake up. So write down the dream as soon as you wake up and go back in the next week or so and read it."

T.L : "Keeping a pad of paper and a pen next to the bed can be helpful, as loved ones many times will appear in a dream. Perhaps there is a message in that dream for you or for someone you know. It is easier to remember these messages if you're able to write them down and share them if needed."

Ginger: "Put a picture of your loved one under your pillow at night or on the nightstand before sleep. Ask for your angels and guides to come in and protect you energetically and to visualize gold light around you so you do not pull in negative or random energies. Look at the picture of your loved one who you want to connect with and ask for that person to come into your dreams to give you a message. In the sleep state, that desperate sense of trying to communicate alleviates and the interaction can take place with ease and comfort. Keep a notebook by the bed to record the dream as they vanish quickly and you will often forget what you experienced. Do this for seven nights straight and watch for signs, such as feathers, pennies, knocks on the wall, scents of familiar perfumes, and songs being played that bring a message from that loved one you want to connect with."

Meditative States

Joanne: "Sit quietly and relax."

Daria: "The practice of meditation allows you to relax and be more open."

Troy: "Meditation, development circles, and prayer are always good ways to help you build your connection to your loved ones."

Dave: "Get into a meditative state of mind, say a prayer asking for protection, and ask to connect to the person you wish. After the prayer, go into or stay in the relaxed state of mind (which takes practice). Then state either aloud or in your mind that you wish to speak to [blank], and ask the person to please give you a message if they can. Then the hard part: Sit silently and wait for 5-15 minutes or until you get your answer. If nothing happens, do not give up! This may take a couple of tries over a week or two. Most people ask for 'something' but then never sit and wait for an answer. Some may get an answer or a message on the first try. I remember trying this with a family member, and the first time I received a very detailed message with instructions to tell my grandmother something."

Daydreams

We've already heard (read) several of the WCRMs liken the communication state to daydreaming. Here are T.L.'s and Daria's suggestions for using daydreams to receive communication.

T.L.: "Be open to messages you receive in your waking state."

Daria: "It's simple; pay attention to your daydreams! One of the most potent tools to connect with deceased loved ones is in daydreams. When you daydream, you raise your consciousness to a higher vibrational plane where you have access to messages and inspiration. This is a great example of how daydreams work: My friend Nancy was walking along and found herself daydreaming about her father who had died when she was a teenager. She wasn't consciously thinking of him; he just popped in. She headed into an antique store looking for Christmas gifts and realized that the song playing in the background was her dad's favorite. She walked to the back of the store and spotted a silver Christmas tree just like the one her mom put up each year which put a smile on her face as her mom had also passed on. Under the tree was a vintage ash tray and on it was printed JOE '28 which was her dad's name and the year he was born. She knew, without a doubt, that both of her parents were still celebrating the holiday season with her. And since she paid attention to her daydream, she was able to connect with her loved ones."

Pets

Joanne: "Our pets can communicate with us as well. When my dog Corky passed away, I would find nickels in her bed or in remote places near her toys. Our beloved pets are also a part of our family, and they stay connected to us through the energy of love, just as our loved ones do. Love is energy and energy cannot be destroyed, but only transformed. Their love has simply transformed."

6

Specific Suggestions

If you skipped right to this chapter, welcome! The WCRMs had many specific suggestions about optimizing your experiences of communication; specifically: be open, release your expectations, ask for contact, be patient, practice, acknowledge contact, and document your experiences.

Be Open

T.L. noted the variety of examples of "how communication between ourselves and those we have lost can occur, but believing we are able to communicate with them and then being open/receptive to the experience is more times than not the key to having that dialogue." Dave echoed that sentiment with the advice: "Be open and receptive. Listen."

Nancy: "The dead don't always come in as birds or butterflies or show themselves full body to us. Many times they show up in their own personalized way. Open yourself up to possibilities."

Carrie: "Be open to the experience, open to the love. Once people learn to open to the message instead of trying to control the message, they get all kinds of unexpected, amazing things. I had a mother call me distraught one Christmas morning. She had lost her 8-year-old son earlier that same year when he was hit by a car. She had barely been out of bed since his death. We talked and he gave her some great validation that he was there, but she just kept saying, 'I need to know for myself.' So, he told me that he had left something outside of the door for her. I told her to walk outside, that he had left a gift for her. She went right outside the door where he said it would be and lying in the snow was the back of a lawn chair that had been bent into the shape of a heart with the eternity symbol at the top. The design couldn't have been made naturally. She keeps it next to her bed and got it as a tattoo. It allowed her to crawl out of bed and go back to work because she knew their love was eternal."

Let Go of Your Expectations

T.L.: "Release expectations. Simply allow things to unfold naturally without forcing anything."

Troy: "Let go of your expectation or attachment to how you want it to look and feel."

Carrie: "Don't set parameters on it. Let it unfold naturally."

Joanne: "One must be willing to let go of expectations and trust in divine timing."

Daria: "Put your preconceived notions aside and open yourself up to the subtleties of how energetic messages arrive. One of my favorite things to say is, 'You are looking for fireworks when you should be looking for fireflies!' Sometimes, we expect to be hit over the head with a communication when they can be very subtle. Leave your expectations behind, and instead set strong intentions of how you would like to receive messages. When my grandmother passed on, my first thoughts were, 'Please don't appear to me; I will freak out!' So instead, I remained open and realized one day that the woman sitting next to me in church smelled just like my grandmother's favorite face cream. So I said, 'OK, thank you, Gram, from now on when you want to get my attention that will be your sign.' And it has worked beautifully ever since."

Ask for Contact

Carrie: "Simply ask for a sign that the person is there with you."

Daria: "Set your intention to experience communication. Like attracts like."

Dave: "If you're asking for a 'sign,' you may give them some ways to let you know or just ask them to find a way to let you know it's them. They are very clever at finding the way. Then give them about a week to show you."

Joanne: "I feel that the energy of intention and love help the process of evidential spirit communication to effectively take place. Put the intention out to spirit and ask loved ones to come and talk to you or give you a sign. Thinking of our loved ones passed creates a telepathic communication. Spirit loved ones can also sense the awareness that you want to connect with them. I do feel that when someone is prompted, especially impromptu, to have a mediumistic reading, their loved one is looking to communicate with them."

Eliza: "Ask your loved one to give you a sign or message. For instance, I asked my deceased father to let me know he did not forget my birthday a few years ago with a sign. So the day after my birthday (I tend to celebrate for weeks), I was sitting at a random stop light and a helium-filled Mylar balloon appeared from the side of the road then flew across my windshield; the balloon had written on it 'For You.' I believe that was my father's way of acknowledging my birthday since I had never seen a balloon float directly in front of me. So if you ask, you will get a sign of some kind and you will know it is from your loved one."

Be Patient

Troy: "Have patience in the process. Spirit communication could come immediately or over time or both."

Daria: "We have gotten used to instant gratification and information on demand. It can take patience to develop new ways to communicate with loved ones and figure out what works for everyone concerned."

Dave: "If you don't feel them or get an instant response to your request, give it time. People ask to hear from the loved one and then go about their business without even taking a moment to pause and meditate. If you're asking for them to communicate with you, ask, then meditate and truly listen."

Joanne: "Be patient. The spirit world does not have a time-space continuum. Your loved one may not come through to you the night that you ask for a dream visit. This is not uncommon. Sometimes it can be days, weeks, or months."

Traci: "The time of those in spirit is not necessarily on the same schedule, calendar, or time clock as ours on this side. We certainly can ask for what we want and need, use miscellaneous techniques, etc., but when it comes down to it, we don't control their communication; they do."

Carrie: "Sometimes our pain gets in the way of us being able to experience the love of someone who we believe we have lost to death. Heal and only try when you feel ready."

Practice

Daria: "As with everything else in life, the more you practice a skill, the more skilled at it you become."

Dave: "If it does not happen the first time, try again. Unless you're already a medium, this may be new to you and you may need to practice opening up for a while before it happens."

Joanne: "Like anything else, learn about it. Read books and take workshops or classes."

Nancy: "Authentic mediums have spent many years perfecting their skills in communicating with the other side. While they may have been born 'talented' in receiving information, most mediums have had to hone their skills by studying books, going to workshops, joining development circles. Many professional mediums have had training to enhance their abilities and continue to study. Yet, in spite of all the experience and all the studying, it sometimes happens that mediums cannot make contact with a specific dead person. You can practice honing your skills and can learn to use your own abilities and will become more sensitive and observant of possible communication. Keep trying. Be patient and understand that if you are open, observant, and study communication with the dead, communication may happen for you."

Traci: "I have to work very hard to do what I do, I've been doing it for decades, and I am still learning. Doing what I do can be equated to learning a foreign language."

Troy: "It takes time. Reading one book or taking one workshop doesn't turn you into a medium. It's like any muscle: if you want to strengthen it, you need to exercise it. Mediumship is a muscle you exercise over years and throughout your life. The learning never stops."

Acknowledge Contact

Ankhasha: "Speak to your loved one and acknowledge every sign that comes from them."

Ginger: "Always trust what you are experiencing. Without the trust, the doubts filter in and messages will be missed."

Nancy: "Educate yourself in the various ways to recognize contact when it is made. If you feel you have had a visitation, trust your heart. You just made contact."

Traci: "If you think you've received a message or sign from your loved one, then know you have. Believe it! Messages are not always a vavoom sign like a full manifestation of the deceased; I don't even see that often. I do not know why a woman who dreams in detail of her deceased husband sitting on the bed stroking her hair questions whether this was really him. Even when she describes feeling this and noting that the sheets felt warm where he sat when she awoke even though her own body was on the opposite side of the bed, she may be uncertain if this was really him. Yes, that was a real experience! I encourage you to say, 'Thank you,' and 'Come again!'"

Document Your Experiences

Ankhasha: "Journal about the signs, feelings, and dream messages you experience."

Troy: "Our loved ones are reaching out to us and we do receive the communication; sometimes it is stronger than other times and also there are times when long periods of time go between communications. It's important to keep a spirit communications journal to keep track of your experiences to remind you of when they occurred, how they occurred, and what made them so profound."

Part 3:

Absence of Communication

Question:

Why might it be that someone has not heard from their loved one when they want to?

The WCRMs could definitely empathize with people wondering why they hadn't heard from their loved ones. Traci revealed, "A number of people have scheduled with me because they are frustrated in that they believe they are not connecting with their loved one and/or their loved one is not connecting with them. They wonder what they are doing wrong, if their loved one is angry with them, or worse, is not 'there/here' and gone forever."

Nancy noted that sitters often ask questions like "Why won't they come through for me?" and "Don't they know how much I want to talk to them?" She wrote, "Many feel forsaken when their loved one doesn't communicate with them. They feel like it is their fault they are not hearing from the person they loved so dearly. So on top of missing the one they lost, many people feel they have done something wrong when they can't connect."

> *The absence of evidence
> is not the evidence of absence.*
> −astronomer Martin Rees

As inferred from the quotation above, just because you have not experienced communication does not mean that your loved one has not attempted contact.

However, it is not a simple phone call. Just as we reviewed some ways that communication can be optimized in Chapter 6, there are many factors that can impede contact as well.

It is important to remember that there are ostensibly two people involved in any after-death communication (ADC) experience: the deceased communicator and the living experiencer. Issues on either person's side may hamper communication. This is true for mediumship readings as well as direct contact. Joanne reported, "It could be that you may not be ready for your loved one to communicate or your loved one may not be ready to communicate with you." In the following chapters, the WCRMs discuss factors on the part of the discarnate communicator (Chapter 7) and on the part of the living experiencer (Chapter 8) that may be affecting communication or its absence.

❦ 7 ❧

Discarnate Communicator Factors

I often remind people: If your loved one's death was a shock for you, just imagine what it was like for them! Being dead probably takes a lot of getting used to and expecting communication on our terms simply isn't fair.

The WCRMs discussed how factors on the part of the discarnate communicator may hinder or entirely prevent communication. These include: they are getting used to being dead; it takes time and effort to learn to communicate with the living; they are experiencing new wonders; they are busy with healing, work, or other tasks; they have moved on to a 'higher plane of existence;' and for reasons we (frustratingly) may never know, it's just not the right time. In addition, Nancy noted that, "Not all dead people want to communicate with the living."

Below are the WCRMs' specific responses to this Why question.

Traci: "Many people who pass may be in a learning mode, learning new things and integrating into a new community which is often celebratory (I would hope that this is so for my loved ones!). Sometimes I use the illustration that the other side may be similar to an 18-year-old going away to college. Living away from home for the first time and enmeshed in a new lifestyle, meeting new friends, and integrating, they forget to call home! Life may be like this for your loved one; rejoice in the joy, freedom, peace, and state of good health they are in."

Daria: "When someone leaves the physical dimension, they have to learn to communicate with energy. Just because a soul is in spirit doesn't mean they are going to be great at it; sometimes it is a learning process. When we understand this, we can give our loved ones a chance to hone their skills. And then there is simply soul evolution. Not every soul wishes to stay connected to this plane. They have 'completed their mission' and are free to move to higher frequencies. This doesn't mean those still here aren't loved. It is just in my work I have connected with souls who have evolved to be a part of various soul collectives who work with beings of higher consciousness. We are, all of us, ultimately completing this journey of duality and separation and striving to reunite to the source, and we will meet again."

Nancy: "The dead do not always communicate with the living. This is a hard fact to accept to those of us left behind on earth. We long to keep them near us. We don't want to let go. It hurts us too bad. But, here's the thing: We have to think about them, too. What do they want? What is best for them? Sometimes the dead cross over and they go directly on to opportunities waiting for them there. They suddenly have a wonderful new life and they become immersed in the wonder of it all. They get busy in the adventure of learning and doing and exploring. Everything is so beautiful that they become mesmerized in the unimaginable radiance of it all. Maybe for the first time ever in their existence, they feel free, unencumbered by anything. So they explore. They roam. They are deliciously happy and absorbed with 'life' in heaven. So if you don't hear from them, don't think they don't love you or care about you anymore. Accept that they are experiencing something magnificent. Be happy for them. You may hear from them later or you may have to wait to see them until you cross over yourself. Either way, transcendence is a good, good thing."

T.L.: "It can be that it isn't the right time for the discarnate to make contact or the discarnate has moved on already."

Eliza: "Sometimes if the loved one passed and there were issues of guilt or unresolved conflicts, it is harder to connect. The loved one might also be 'working' on assisting someone else."

Carrie: "Once I had a third party discarnate come through on behalf of the spirit someone wanted to talk to and say that the spirit was tied up and couldn't come at this time. Oddly, this was the same message the couple had gotten from another psychic medium a week earlier. So, it turned out to be very validating for them."

Dave shared some personal experiences to demonstrate discarnate issues that may affect communication: "Recently my cousin committed suicide. When my mom was told of his passing, she tried to tune in to him (she is a medium also); he appeared with his head down and he said no (meaning he didn't want to talk). 'I just didn't think it through.' So he did not want to communicate; it was too soon, and he was very upset at what he had just done. This is very common. Many people are still in depression and getting needed help on the other side. They may not be ready to communicate while in that state of mind. Sometimes it is that they will stay 'around' for a while and then move on to another plane. This does not mean they cannot come back and visit or communicate, but they may also be doing other projects that are taking up their time. I had a friend who passed and stayed around for a while then he came to another mutual friend of ours and told him, 'I'm leaving now, you would too' in a very audible voice. I have not felt him around like I could for a period of time, only on occasion will he come through, and usually at my request to communicate with him."

❦ 8 ❧

Living Experiencer Factors

The WCRMs shared many issues that may be affecting communication on your end. Daria listed unrealistic expectations, fear, grief, and disappointment as "the biggest road blocks to receiving messages." Below, the WCRMs discuss these and other elements including not recognizing communication when it happens, beliefs or doubts (often subconscious) that prevent contact, general health status, and wanting it too badly.

Unrealistic Expectations Related to Media Portrayals of Contact

A number of the WCRMs lamented how fictional and highly-edited portrayals of mediumship on TV and in the movies have given the public unrealistic expectations about the subtle nature of communication.

Ankhasha: "I find that the media has really contributed to the high level of unrealistic expectations that some of the bereaved have when it comes to hearing from their loved ones.

Although many of the shows have given an opportunity for mediums to reach a wider audience, the time constraints of a television show or a movie require that 'reality' is edited and set up to reflect a medium as in constant communication with the dead. They appear to be able to communicate clear, exact messages as reliable as a telephone line."

Traci: "We often talk about television mediums and that it looks really easy on television to connect because the 22 minutes of a 30-minute show is often edited down from 5-8 hours of taping. I believe that the media can set up people for feeling like they are failing at communication. Media portrayals can create grandiose expectations for people. In other words, if one is expecting a message from their loved one to come in the form of a full manifestation, that is likely unrealistic."

Nancy: "Television, books, and movies make it look like mediumship is easy to do. But the viewer should remember they are observing someone who is practiced and who has spent many years developing their 'talent.' Often media makes it appear easy and an event to be expected. This builds up huge expectations on the public for performance. As a result, many times people feel they are not loved when their loved one doesn't contact them. People also think something is 'wrong with me' if they can't communicate with spirit. So they become very frustrated and angry with themselves. Sometimes they even get angry with spirit."

Joanne: "Spirit communication is not on demand as so often portrayed in Hollywood movies such as *Ghost*."

Troy: "Often times people think the communication will be some profound experience like they've seen in the movies or on TV, but communication is often a very subtle experience."

So try to remember that---like is true for so many things---it's not like what you've seen on TV.

Not Recognizing Communication

With unrealistic expectations comes the potential inability to experience and recognize subtle forms of communication. It seems like that would be frustrating for a discarnate attempting to communicate.

Traci: "What I suspect is that many people are missing messages and signs being sent from those on the other side. They may well be hearing from the loved one, but are not recognizing it."

T.L.: "Don't disregard things that seem too simple, as if the message should come up and hit you right upside the head. Sometimes subtle is just as validating and important as the more obvious messages you might receive."

Daria: "People often ignore the messages that come to them because they are used to dealing with physicality and not pure energy."

Joanne: "To 'hear' from a loved is not always an audible experience. A loved one is not going to suddenly appear in full, out of thin air to say, 'Hello, I am here.' It would be quite startling, even to most mediums! It's about awareness. If we are not aware, we might be missing the signs. Spirit has many different ways and forms to communicate."

Troy: "It's a subtle experience so it might not be through hearing; it might be through a sign or dream."

Ankhasha: "I have had clients who have attempted contact with their loved ones for years to no avail, and yet they remain assured that they continue on the other side. And I have had clients who receive amazing validating evidence in great detail and yet remain unconvinced. They may see multiple signs that they are being contacted including dream visitations and still they wonder, 'When will I really know they are there?'"

Beliefs and Doubts

Nancy: "Maybe, deep down, they just don't really believe it is possible to communicate with a dead person. So in reality, they are not really open to receiving information. Even if they are contacted they may dismiss it as being not real or only their imagination."

Dave: "Many times it is you, the sitter, who is having a hard time opening up. It may be that you're not very intuitive naturally or open minded (too skeptical) to see when there are flagrant signs right in front of you. For example, I say something very specific and they still don't get it until the light bulb goes off, because they expected me to say something else, but the thing I brought up was specific and detailed enough that they knew it was the crossed-over loved one."

T.L.: "The mere idea that someone could even do this---reach out and 'connect' with a loved one---on a conscious level seems, well, odd, for a lot of people. I hear often from clients that they struggle with believing they truly have any mediumistic abilities, that they 'feel silly,' they think they look or sound 'ridiculous,' or that they are 'making things up' when trying to communicate with a loved one. They feel if they receive any signs or messages, it's more from a place of wishful thinking that a loved one is reaching out to them than actually knowing, concretely, they are truly having the experience. So, they question themselves a lot, they second guess their abilities and ultimately wind up getting in their own way which is often why communication is hampered. Personal views such as religious beliefs can also make it intimidating or frightening for someone to try and make contact."

Fear

I make my husband quickly scroll past the Horror titles on Netflix because the images are too scary for me, so when I started performing research with mediums, I thought I would turn into a quaking puddle every day. But just like the media has misrepresented ADCs, it has also given the dead an undeserved bad rap: I soon learned that they are, in fact, not scary. The WCRMs reported that those feelings I had may be affecting contact in others.

Nancy: "Some people are afraid of contact with the dead and unintentionally sabotage the communication. So actually making contact, even if it is with someone they love, seems frightening. So they shut down."

Daria: "Fear can block communication. A loved one may try to get our attention by making noises or affecting electricity and it is scaring us. The more we try to ignore it, the more energy they put into it. I find saying out loud, 'This is not working and we have to try another way to communicate' often works. They really are just saying, 'Pay attention.'"

Grief

The WCRMs had a lot to say about how grief can impede communication.

T.L.: "While it sounds counterintuitive to not be able to connect with a discarnate when we are grieving, as that's the whole idea of wanting to connect in the first place, I believe the intensity of the grief, the depth of one's pain, creates the inability to 'feel' or 'sense' beyond what you are currently experiencing emotionally. Hence, your senses become too overwhelmed, too blocked because often times it's the emotions which keep those hoping to connect anchored in their own pain rather than open to communicating with a departed loved one."

Dave: "One reason some people don't get communication is that they aren't ready. The spirit world knows this. They either can't come in due to the overwhelming emotions you are feeling or they just choose not to come until you are stable enough to receive a message."

Troy: "Depending on the level of grief you are experiencing, you might not be 'open' to the communication in the same way. Time heals and individuals should recognize that if their grief is incredibly deep they should allow themselves that journey first and foremost."

Nancy: "Maybe your own grief and strong feelings for the dead person make it hard for them to use their abilities. (Many professional mediums have difficulty reading for their family and people they love.) Strong, personal emotions get in the way of receiving communication and information from the other side."

Daria: "Often I find that people who are still deeply grieving think they want to communicate, but are really afraid that if they do, the 'flood gates' of tears will open and they won't be able to deal with the emotions. It is also difficult for higher vibrational beings to pass through that wall of grief and so they wait until there has been time to heal. Then the deceased will often come through with humor, as if to say, 'All is well.'"

Eliza and Carrie shared their own experiences of grief preventing communication.

Eliza: "The person who says they want to feel or hear their loved one may truly not be ready to do so due to their grief or inability to forgive. I personally had this experience with my brother who passed 30 years ago and to whom I was very close. Right after he passed, I had multiple visitations and dreams but it had been many years since I had felt his presence so I had a reading with a medium who said my brother was far away on a planet and was very busy. (Sometimes, as mediums, it is difficult to do your own readings). She said he and I will be together on this planet someday when I pass. She confirmed that he did visit me a long time ago and that he loved me very much. I believe I really wanted to connect finally with my brother but prior to that time I was hurting so deeply from his passing that I emotionally blocked his attempts to connect with me."

Carrie: "I find that very frequently people get in their own way of communication with their loved ones. The best example I can give is my own experience. I talk to dead people for a living, yet after losing three children of my own I was angry with God because I could not feel my own children. I sat at the end of a very long day of doing readings and I said to God, 'I refuse to do this work for you any longer if you will not let me talk to my own children,' blaming God for my inability to connect. At that very moment a man walked past me and stopped. He asked, 'Do you have three children on the other side?' I said, 'Yes.' He said, 'You look too young to have three children on the other side. They are always with you.' Then he kept walking as if God was saying, 'They are always with you. It is you who can't feel them.' And I couldn't because my own emotions got in the way. I find very often that people who are looking for a sign and don't get one from their loved ones may just be missing their communication through their own sorrow."

Health

T.L. thought that overall health may be having an impact in some cases, including "a person's lifestyle and the foods they eat (i.e., too much sugar or processed foods) which can create a feeling of fogginess or depletion in their body/mind hindering their ability to connect."

Kim noted that in her own mediumship readings, "I make sure I keep hydrated during all sessions."

Trying Too Hard

Troy: "You might be forcing the communication rather than just allowing it."

Ginger: "The extreme wanting of it often times cancels out the interaction and keeps the person who is trying so hard in a state of negativity and frustration."

You've probably heard stories about couples who try and try to conceive and only once they have adopted a child do they get pregnant. Or you may be familiar with the notion that you won't find love until you stop looking for it. Perhaps communication is like that: you may only be able to achieve it once you stop trying.

Who Knows?

The WCRMs also discussed how in some cases we'll never know why communication doesn't happen. It's important to remember that communication between the living and the deceased still falls into the realm of the mysterious: no one has all the answers.

Traci: "There is so much more that I do not know. I do not know why Aunt Minnie shows up first in a reading when the woman I'm reading for either never knew Aunt Minnie or detested her. But I know that Aunt Minnie may be appearing to help introduce others that are there waiting to come through. Therefore, it is possible that signs are coming through other deceased people/animals that may contribute to them missing who it is they wish to hear from."

Nancy: "Even professional, authentic mediums do not always make contact even after years of studying. We don't always understand why either. Like you, we just have to keep searching."

❦ 9 ❦

Still Not Receiving Communication?

We all can't paint masterpieces and we all can't play in the NBA. Such is the bell-curve nature of life.

Recently, the fortune in my tiny tin of delicious sweetriot cacao nibs was a quotation from Maya Angelou: "If you don't like something, change it. If you can't change it, change your attitude." You may not be a medium; you may not even have any talent for recognizing signs from the deceased. But that is just a condition of your existence; it doesn't have to mandate its quality. You can still enjoy your memories of your loved one and the memories you make with the loved ones still here.

To conclude this volume, here is Nancy's advice for you: "You should forgive yourself if you do not make contact. It is not your fault. Forgive your loved one, too. They may not know how yet. Your loved one still loves you even though you haven't heard from them. It may be that you are not supposed to be making contact. Just relax and be you. You are still here. And while your loved one may not physically be here with you, you are still loved. You are both souls, just residing in different places. Energy cannot be destroyed. It just changes form. Love is still love no matter what dimension it resides in. Like you, the dead are trying to understand their world, who they are, where they are. You are wonderful and so is your loved one. In the end, love is what matters."

From the Mouths of Mediums Vol. 1 Contributors

Ankhasha Amenti is an investigator for the Office of Paranormal Investigations and received the Providence Hospice of Seattle Foundation Award of Distinction. She loves the rain and lives in Sammamish, Washington, with her husband, two cats, and a Velcro Border Collie.
http://ankhasha.com/

Traci Bray, BS, MA, has been reading as a medium and psychic since the mid-1980s. She travels widely reading for families, groups, and individuals, as well as from her home in Kansas City, Missouri. Pets and gardening are her passions.
http://tracibray.com

Dave Campbell, author of *Forensic Astrology: Solving Crimes with Astrology*, lives in Phoenix, Arizona. He and his partner own The Astrology Store, a metaphysical bookstore in Glendale and have three Chihuahuas, a talking African Grey parrot, and two owl finches.
http://theastrologystore.com/

Carrie D. Cox doesn't remember a time when she couldn't connect with the dead and has been working as a professional psychic and medium for over 20 years. She lives in Burlington, Kentucky.
http://carriedcox.com/

Joanne Gerber is internationally recognized for her natural ability to bring through information from loved ones passed on. She lives in the Boston, Massachusetts, area with her partner and Louis, a West Highland Terrier. Joanne enjoys reading, cooking, and traveling.
http://www.joannegerber.com/

Daria Justyn spent her life cultivating her ability to channel messages from angels, guides, ascended masters, and departed souls. She lives in her favorite 'city of ghosts,' Charleston, South Carolina, and is VP of The Harmony Fund animal rescue charity.
http://www.dariajustyn.com

Nancy Marlowe, author of *Let Your Soul Shine: Thoughts of a Psychic Medium*, lives by the ocean in San Diego, California. She loves solving mysteries. Nancy is a forensic medium, evidential medium, personality profiler, and psychic detective.
http://nancymarlowe.com/

Tracy Lee (T.L.) Nash, an Interfaith Minister in Southern California, founded the metaphysical ministry Within The Light and The Good Grief

Circle and is happily married with four children. T.L. can also whinny just like a horse and they always whinny back!
http://www.withinthelight.com/

Troy Parkinson, author of *Bridge to the Afterlife*, has studied mediumship for over 15 years and is also a professional media producer. He resides in St. Paul, Minnesota, with his wife and three children.

Ginger Quinlan, author *Scents of the Soul* and *Shining Through the Shadows*, lives in Atlanta, Georgia. She mentors budding mediums, does energy work, and is a Master Herbalist who loves to play in the dirt in her off time.
http://www.gingerquinlan.com/

Eliza Rey, author of *Daysee, The Delinquent Angel* and the sequel *Last Wishes*, developed the program Intuitive Tools 4 Kids. Eliza has had mediumistic abilities since she herself was a young child. She lives in the Phoenix, Arizona, area.
http://elizarey.com/

Kim Russo (a.k.a. The Happy Medium), host of the TV show "The Haunting Of...," loves spending quality time with her husband of 27 years and her three sons on Long Island, New York, where she lives.
http://www.kimthehappymedium.com/

INVESTIGATING MEDIUMS

Julie Beischel, PhD

Academic Negligence or Economic Reality?

from *Winds of Change*, Spring 2009,
Vol. 2, No. 1, pp. 10-11

In January 2009, Victor Zammit, an Australian attorney and "Lawyer for the Afterlife" (www.victorzammit.com), posted a commentary in his "Friday Afterlife Report" titled "Why the Colossal Negligence by Universities?" which asserted that "the paranormal can be objectively measured" and asked "Why aren't universities doing controlled research into the greatest discovery in human history? Is it materialist stubbornness? Materialist denial? Materialist academic gross negligence?" (January 23, 2009).

As Survival Researchers ourselves, Mark Boccuzzi and I felt it was important to address in a letter to Victor the economic reality of scientific research and its impact on the choices scientists make. Below is that letter which Victor quoted in a subsequent Report and posted on his site in full (www.victorzammit.com/articles/Windbridge.htm).

Julie Beischel, PhD

Hi Victor,

We hope you and Wendy are well. Mark and I wanted to thank you for your continued support of The Windbridge Institute and for always listing Windbridge in your weekly Afterlife Report. We also wanted to respond to your previous commentaries about university research.

While it is true that only a very small percentage of universities (or other research institutions) actually perform paranormal research, what is preventing more investigators from engaging in such research has much more to do with money than with "negligence" on the part of the institutions.

It is important to note that universities do not, for the most part, fund research. They provide space and the basics like electricity for laboratories, but the researchers themselves are responsible for acquiring the funding needed to support their studies. This funding is used to pay the salaries of the research staff as well as to pay for equipment, computers, copiers, paper, printer ink, books, journal subscriptions, society memberships, travel for academic conferences, etc., etc., etc. Researchers most often acquire funding through grants offered by--in the US--the federal government or certain private organizations. Rarely, funding is provided by generous individuals interested in certain types of research.

Usually, however, individual donations go more to support the university infrastructure (for example, new buildings) than actual research studies. In addition, at a university, close to half of the acquired funding regardless of its source goes to the institution for overhead costs and does not directly support research.

This paradigm--again, at least in the US--forces researchers in any and all scientific fields to choose between the studies they would like to do and the studies for which they can acquire funding. Thus, it is the "negligence" of the funding organizations and not the universities per se that most often prevents parapsychological research from being done. Granted, the stigma of performing fringe parapsychological or psi research ("the boo taboo' as Dr. Dean Radin calls it) does its share of preventing researchers from embracing these fields, but if there was money in them, it is only logical that a lot more individuals would go for it anyway. No one should be blamed for wanting a job that allows for the expenses of shelter and food.

The reality is that there is very, very, VERY little funding available to perform parapsychological research. This in turn also makes it difficult to "waste" funding on systematic replication studies that would solidify the reality of the phenomena being studied. Researchers and funding organizations alike prefer to use the limited resources to investigate new questions or use new methods. This, however, prevents studies from being replicated in order to increase the amount of evidence for any particular phenomenon.

The current global economic crisis is making matters even worse. A number of groups in the parapsychology community as well as in other fields are not able to offer the grants this year that they usually do. In the November 7th, 2008, issue of the journal Science, *the "News of the Week" section included an article by Jennifer Couzin about the impact of current economic woes on the funding of scientific research. "Among the first to feel the slowdown," writes Couzin, "are charitable foundations and other philanthropies, which provide billions of dollars in funding to scientists each year, including support for innovative, risky research that the government may be reluctant to back." Scientists all over the US are losing their jobs because even philanthropic organizations which rely on endowment income are working with limited support. At Windbridge, we strongly rely on the support of our members and donors in order to perform "innovative, risky research that the government may be reluctant to back."*

At Windbridge, we were very honored to receive funding this year from what may be considered the largest parapsychological research grant available in the world: The Bial Foundation, associated with a major pharmaceutical company of the same name in Portugal, offers a biannual grant for research in parapsychology and psychophysiology.

We are so grateful for this support, but to put it in perspective, the yearly funding from a Bial research grant is a mere fraction of the yearly funding provided by US federal grants from, for example, the National Institutes of Health (NIH) or National Science Foundation (NSF) for mainstream scientific research studies. Our project--an objective analysis of mediums' abilities that is a replication of a previous study I performed while at a university--requires four researchers as well as equipment, office supplies, etc. and even with little to no overhead costs at Windbridge, the Bial funding does not go as far as we would like and there are few other options available for us to support that and the other studies we hope to do this year.

So while ignorance and materialist denial on the part of university researchers and administration definitely play a part in the absence of psi research at those institutions, the real issue is most likely an economic one.

Best wishes,
Julie and Mark

Julie Beischel, PhD, Director of Research
Mark Boccuzzi, Director of Operations
The Windbridge Institute
 for Applied Research in Human Potential
www.windbridge.org

Julie Beischel, PhD

Further Resources

The Windbridge Institute

Launched in 2008, the Windbridge Institute, LLC, is dedicated to conducting world-class research on phenomena currently unexplained within traditional scientific disciplines. Our primary focus is on applied research with the goal of developing and distributing information, services, and technologies that allow people to reach their full potential so they can live happier, healthier, and more fulfilling lives. For more information about Windbridge Institute research, publications, news, events, media, investigators, and advisors; for a list of our Windbridge Certified Research Mediums; or to join our email list, please visit: http://www.windbridge.org

For ways to get involved and stay connected, visit: http://www.windbridge.org/get-involved

Dean Radin: Selected Peer-Reviewed Publications on Psi Research

This site contains a selected list of downloadable peer-reviewed journal articles on psi (psychic) phenomena, most published in the 21st century. There are also some important papers of historical interest and other resources. A comprehensive list would run into thousands of articles. Commonly

heard critiques about psi, such as "these phenomena are impossible," or "there's no valid scientific evidence," or "the results are all due to fraud," have been soundly rejected for many decades.
http://deanradin.com/evidence/evidence.htm

American Center for the Integration of Spiritually Transformative Experiences

Rather than attempt to theorize how near-death, out-of-body, kundalini, spiritually emergent, after-death, mystical, or other experiences occur, the American Center for the Integration of Spiritually Transformative Experiences (ACISTE, pronounced 'assist') works to research, educate, and support people who have had these experiences. ACISTE was "established to support people who have had spiritually transformative experiences (STEs)." They offer: competency guidelines for professionals working with clients who may present issues related to their STEs; education, training, and certification to professionals who want to provide support for experiencers; a directory of professionals certified by ACISTE; and free online and regional support and discussion groups for individuals who've had STEs.
http://aciste.org/

Forever Family Foundation

The Forever Family Foundation's missions include establishing the existence of the continuity of the family, even though a member has left the physical world; stimulating thought among the curious, those questioning their relationship to the universe, and people who are looking for explanations of certain phenomena; financially supporting the continued research into survival of consciousness and afterlife science; and providing a forum where individuals and families who have suffered the loss of a loved one can turn for support, information, and hope through state-of-the-art information and services provided by ongoing research into the survival of consciousness and afterlife science.
http://www.foreverfamilyfoundation.org

The Society for Scientific Exploration (SSE)

The SSE is a multi-disciplinary professional organization of scientists and other scholars committed to the rigorous study of unusual and unexplained phenomena that cross traditional scientific boundaries and may be ignored or inadequately studied within mainstream science. The SSE publishes a peer-reviewed journal, the *Journal of Scientific Exploration* (JSE), and holds annual scientific meetings in the USA and periodic meetings in Europe. Topics addressed in its journal and in its regular meetings cover a wide spectrum,

ranging from real or apparent anomalies in well-established areas of science to paradoxical phenomena that belong to no established discipline.
http://www.scientificexploration.org/

Rhine Research Center

For the past 70 years, the Rhine Center has been researching and studying the experimental science of parapsychology. Now in the 21st century, the Rhine Center continues the mission and work of its founder J.B. Rhine with a broadened scope directed deeper into the Study of Consciousness.
http://www.rhine.org/

Society for Psychical Research (SPR)

The SPR, located in the United Kingdom, was the first organization established to examine allegedly paranormal phenomena using scientific principles. Their aim is to learn more about events and abilities commonly described as "psychic" or "paranormal" by supporting research, sharing information and encouraging debate. SPR members, who come from all over the world and represent a variety of academic and professional interests, have access to a comprehensive online library as well as the full content of SPR's publications from the very first issues to the present day and a number of out-of-print classic books and materials housed at the SPR libraries in London and Cambridge.
http://www.spr.ac.uk/

About the Author

Julie Beischel, PhD, co-founder and Director of Research at the Windbridge Institute, received her doctorate in Pharmacology and Toxicology with a minor in Microbiology and Immunology from the University of Arizona in 2003. She is a member of the Society for Scientific Exploration (SSE) and serves on the scientific advisory boards of the Rhine Research Center and the Forever Family Foundation. Following the suicide of her mother and an evidential mediumship reading, Dr. Beischel forfeited a potentially lucrative career in the pharmaceutical industry to pursue rigorous scientific research with mediums full-time. Her current research interests include examinations of the accuracy and specificity of the information secular, American mediums report as well as their experiences, psychology, and physiology and the potential social applications of mediumship readings. Dr. Beischel is Adjunct Faculty in the School of Psychology and Interdisciplinary Inquiry at Saybrook University and Director of the Survival and Life After Death research department at the World Institute for Scientific Exploration (WISE). Dr. Beischel lives in Tucson, Arizona, with her husband and research partner, Mark Boccuzzi. For more information, visit www.windbridge.org

Julie Beischel, PhD

INVESTIGATING MEDIUMS